THE AMERICAN PEOPLE & THE NATIONAL FORESTS

THE AMERICAN PEOPLE & THE NATIONAL FORESTS

THE FIRST CENTURY OF THE U.S. FOREST SERVICE

Samuel P. Hays

UNIVERSITY OF PITTSBURGH PRESS

Published by the University of Pittsburgh Press, Pittsburgh, Pa., 15260

Copyright © 2009, University of Pittsburgh Press

All rights reserved

Manufactured in the United States of America

Printed on acid-free paper

10 9 8 7 6 5 4 3 2 1

ISBN 13: 978-0-8229-4369-3 (cloth)/978-0-8229-6020-1 (pbk.)

ISBN 10: 0-8229-4369-7 (cloth)/0-8229-6020-6 (pbk.)

A Cataloging-in-Publication Data record is available from the Library of Congress.

FOR

DANNY, KATIE, ROWAN, RUBY, AND SARAH

CONTENTS

Preface		ix
Acknowledgments		xv
1	A Century of Change	1
2	The Silvicultural Imperative, 1891–1920	25
3	Evolution of an Agency Clientele, 1920–1975	55
4	Confronting the Ecological Forest, 1976–2005	106
5	Epilogue: Facing the Future	137
	Notes	145
	Sources	183
	Index	191

PREFACE

THE YEAR 2005 marked the centennial of the U.S. Forest Service. Until February 1, 1905, the nation's forest reserves had been under the jurisdiction of the Department of the Interior. On that date, the Department of Agriculture's Bureau of Forestry assumed control of the forest reserves and the bureau debuted under a new name: the U.S. Forest Service. The one-hundredth anniversary of the agency seemed a propitious time for historians to revisit its history. The passage of a century provides critical perspective with which to view just how much the agency had changed over that period of time.

I was particularly interested in having an up-to-date account of the agency's history that would incorporate material written since I first became interested in forest history. At that time, in 1948, the only useful source about the agency's history was Gifford Pinchot's autobiography, *Breaking New Ground*, published in 1947. Since that time, and especially since 1970, more than a hundred books and innumerable articles have focused directly on important aspects of Forest Service history or have covered the topic in more oblique ways.[1]

For the centennial itself, the agency produced both a DVD, which drew upon many interviews with both agency participants and historians, and a companion book, *The Forest Service and the Greatest Good: A Centennial History*, written by James G. Lewis. When the Forest Service released the DVD at the American Society of Environmental History's annual convention in 2005, the ASEH echoed the federal agency's self-congratulatory tone. The ASEH's newsletter later published a review that praised the DVD as a "rich and satisfying historical narrative." The agency's own version of its century-long history ignored the rich accumulation of scholarship about the national forests and the agency. The DVD and this ASEH review thus made a serious revision of the agency's history a timely project.

The agency's version of its first hundred years did not offer significant additions to or revisions of previous accounts. Its narrative of events up to 1920 seemed only to repeat the perspective of *Breaking New Ground*. A history published nearly sixty years after that book should have contained considerable revision, especially in its treatment of the role of Gifford Pinchot.[2] Further detracting from its historical value, the agency's centennial history devoted very little or no attention to such topics as wildlife, recreation, watersheds, and aesthetics, and historical perspective was nonexistent for the tumultuous years since 1970.[3]

The centennial year seemed like an opportune time to employ a new approach to the history of the Forest Service. The events of the last decades of the twentieth century forcefully suggested an approach that would address the relationship between the agency and society as it evolved during those years. Traditional accounts stressed "legislative" history, a context that brings together laws and agency policy in a legalistic pattern of administrative evolution.[4] But in the years after 1970, the more compelling drama was the way in which that administrative structure was challenged time and again by impulses emanating from society at large. The agency's responses to those impulses constituted major historical shifts. This type of wider historical perspective would put more emphasis on the broader forces impacting the agency and on the innumerable choices it made in response to those external impulses.

This interaction between society and agency was an ever-present context for the agency's history from the 1970s to the beginning of the agency's second century in 2005. Applying that perspective to a history of the national forests all the way back to the first authorization of the forest reserves in 1891 should provide new insights, and that is the goal of this book.

Since the focus of the following chapters is on the relationships between American society and the U.S. Forest Service, they are organized around different facets of those relationships. They emphasize the changing interest of the public in national forest affairs, track the initiatives of the public, professionals and scientists, and agricultural and industrial leaders in their efforts to influence the agency and recount the responses on the part of the

Forest Service. Did innovations in management, for example, arise from within the agency or from outside it? If from the inside, which were expressed more forcefully and which less so? If innovations arose from outside, did the agency respond with support and encouragement or with indifference or opposition? To what extent did external influences modify the initial course set by the agency at its origins? Were agency responses shaped by its traditions of professional training and experience or did it modify those traditions in response to demands that the agency change direction? The relations between society and agency cast the history of the Forest Service in a new and more meaningful light.

Another emphasis in these chapters is the decision making of Forest Service administrators. Many administrative choices reflected not statutes or decisions made by Congress but internal agency affairs. One is struck by the fact that the requirements of statutes such as the Forest Management Act of 1897 played a limited role in the agency's history; the historian must go directly to day-to-day records of agency experience and decisions to recount its history. For example, grazing, which was not identified in the 1897 law as one of the agency's responsibilities, was taken under its purview upon its establishment and for a number of years was its dominant preoccupation. Watershed management, on the other hand, one of the two explicit agency responsibilities identified in the 1897 law, received little management attention from the Forest Service over the years and was continually subordinated to other program objectives; it was not until after World War II that hydrologists became a part of the agency's management personnel.

Numerous situations and decisions clearly illustrate the importance of administrative choices in society-agency relationships. For example, timber sales contracts and grazing permits reflected an ongoing pattern of relationships between the agency and its extractive industry clientele. In a particular instance of decision making, the agency rejected the use of the national forests as wildlife refuges. It set a policy of not harvesting timber in areas frequented or traveled by the public because people might react negatively to the "aesthetics" of vistas altered by logging. The agency supported a law—the Municipal Watershed Act of 1940—that would make it difficult for cities to protest timber harvesting in their watersheds. After World War II, the agency abandoned its policy requiring selective harvesting of trees from timber stands and began to allow clear-cutting. In 1952, the agency decided to reevaluate all primitive, wild, and wilderness areas for possible boundary revision. These and many other decisions were administrative choices, not legislative mandates; they constitute the fundamental objectives one must investigate when tracing the relationships between the agency and American society.[5]

An additional feature of this agency history is its stress on patterns of activity rather than the influence of landmark events. Some historians are attracted to dramatic episodes as the pegs on which to hang their understanding of the past. In the case of the Forest Service, such major episodes include the Denver Public Lands Convention of 1907, in which western leaders blasted Pinchot and the agency; the Pinchot-Ballinger controversy, which led to Pinchot's ouster as chief of the Forest Service; the construction of the Hetch Hetchy dam, an example of the ongoing conflict over aesthetic objectives in resource management; the dramatic fires of 1910 in northwestern Montana; the Monongahela case, in which a federal court declared clear-cutting contrary to the Forest Management Act of 1897; and street demonstrations and sit-ins during controversies over agency management in the last thirty years of the twentieth century. Episodes such as these, which constitute prominent features of the agency-sponsored centennial history, while attractive for their drama, often obscure far more than they illuminate.[6] Much more useful for understanding the Forest Service's broader history are the patterns of administrative decision making and the relationships between society at large and agency personnel.

In a similar fashion, telling the agency's whole story requires far more than simple chronology. A timeline can often hide rather than bring to light the sequences and patterns of historical development. In this book I emphasize trends over longer periods of time rather than chronology, and these are trends that involve patterns of stability or change over the long run. Most frequently they overlap, with new trends emerging while old ones are declining or new ones evolving while old ones persist. The major developments in the history of the Forest Service occurred over the long run, such as the evolving relationships between the agency and the timber industry, the gradual acceptance of recreation, the mixed reactions to wildlife, or the persistent aversion to a formal recognition of natural beauty as an objective in itself and the acceptance of that aesthetic only when it seemed essential to accomplishing other agency objectives.

I take up two strategies to identify patterns of long-term trends in the relationship between society and agency. The first organizes history over the entire century, from 1905 to 2005, by observing several century-long transformations that took place; this is a method of dealing with a big question: "In what fundamental ways did the relationships between society and agency change over those years?" The second approach involves outlining three major time periods, each of which witnessed a new pattern developing in the relationships between society and the agency. Each new trend served as a legacy that helped shape administrative decision making in the succeeding years. In some cases the legacies continued in force, exercising influence over much of the agency's subsequent history.

By the end of the Forest Service's first century, it had accumulated a set of legacies derived from different segments of the past. In their cumulative overlapping, they continued to shape the agency at its centennial. It was with that cumulative inheritance that the agency faced its second century. It is also with that inheritance that the century-long history of the Forest Service can most fruitfully be understood.

ACKNOWLEDGMENTS

THIS thematic history is drawn largely from the many books that have been written about the U.S. Forest Service since 1970. The years between 1970 and 2000 were exciting and dynamic ones for both participants in and observers of that federal agency's affairs, and my own experiences in the management of forests at both the state and the federal government levels have deepened my understanding of the Forest Service's activities. The broad perspective of the Forest Service I present here has thus evolved from a wide range of recent scholarship and personal experience.

I am indebted to many people for providing me with this lifetime of experience. They include officials of the Oregon and California Revested Lands Administration in the 1940s, specialists in natural resources at the Library of Congress and the National Archives in the late 1940s and 1950s, the forest administrators and members of citizen groups with whom I was associated in the 1970s and 1980s, and the archivists at the University of Pittsburgh with whose help I preserved the many documents on forest affairs I collected in those years. A work such as this history synthesizes a great deal of research, many experiences, and a lot of ruminating about it all. Every aspect of that developmental process employs the input of a considerable number of people, each of whom unknowingly has made a contribution to the final writing. Of course, I alone am responsible for the resulting perspective and its many facets. But I offer my sincere thanks to one and all for their participation in the journey.

A special acknowledgment also goes to the University of Pittsburgh Press and to its director, Cynthia Miller, for continuing to publish my books in environmental history—this is the fifth one—and to the succession of editors, indexers, and marketing personnel who have worked on them so effectively. My thanks to all.

THE AMERICAN PEOPLE
& THE NATIONAL FORESTS

1 A CENTURY OF CHANGE

I EMPHASIZE five major developments that occurred in the history of the Forest Service over its first century, from 1905 to 2005, each of which took place against the backdrop of the prior history of the forest reserves that began in 1891, when the first legislation leading to the Forest Service's establishment was passed. These developments were the initial commitment to the forest reserves as publicly owned resources. That effort was soon followed by battles to ward off their privatization. From its earliest years, the agency commenced a long-term campaign to shape private forest management, a task that the federal agency effectively abandoned in the early 1950s and nominally and selectively shifted to the states. Later, citizens from outside the agency attempted to become involved in its affairs, which engendered varied responses from the agency itself. From the 1970s onward, the participation in forest affairs by diverse groups of scientists and professionals accompanied changes in the public context within which the agency carried out its objectives. Finally, the Forest Service has been steadily challenged to develop its management capabilities over the years as its tasks have become more elaborate and complex.

From Private to Public

When Congress passed an omnibus public lands bill in 1891, the legislation contained a brief provision that authorized executive orders to establish forest reserves. This move was one of a number of decisions made by Congress and the president that modified traditional public land objectives. The traditional objective had been to shift control of timber lands to private individuals and corporations. With the new legislation, the focus shifted to retaining and managing forests as public assets. In earlier years, Congress had taken similar action only in special cases, when, for example, it excluded from sale areas that contained timber that was critical for naval shipbuilding. Later in the nineteenth century, however, Congress began to assert public interest in natural resources on a wider scale, establishing public lands for parks, wildlife refuges, and watershed protection, as well as for timber production. Congressional designations of public lands that became forest reserves were the most numerous of these federal actions to assert public interest in natural resources. The western forests were initially called "reserves" with the notion that lands should be "reserved" from sale to private parties and retained as public assets to be managed by public agencies. In the East, a policy of acquisition as well as retention of forest land was adopted, leading over the years to the addition of almost 25 million acres to the national forests.

The primary implication of this significant shift in public land policy was the belief that private owners could not be relied upon to accomplish a wide range of public objectives. If the nation wanted undeveloped park areas for the enjoyment of its citizens, it could not depend on private enterprise to retain the land's natural qualities for that purpose. If the nation wanted abundant game and wildlife, which legally belonged to the public, it could not rely on private enterprise to provide either; the economic urge to sell wildlife in the market would be powerful. If the nation wanted to preserve archeological and historic sites, private enterprise could not be relied upon for the same reason. If the nation wished to guarantee future supplies of timber, the drive for short-term economic profit could not guarantee restoration of cut-over lands because this activity required a longer-term commitment of investment with lower returns than private enterprise could tolerate.

Behind the creation of all such "reserves," particularly the forest reserves created under the 1891 act, was the notion of *public* enterprise, which could be carried out by public ownership and management. By establishing these reserves, Congress made permanent public assets out of the public lands that had previously been heavily subject to privatization. The U.S. Forest Service was constituted to manage the most extensive of these public assets.

The nation's forests presented a distinctive example of the relationship between private and public ownership of land. Those who exploited the standing timber of the nation's forests had migrated from New England to New York and Pennsylvania, then to the upper Midwest around the Great Lakes, then to the South, and finally to the Pacific Northwest. These lumbermen were prone to "cut and run," that is, to cut down the existing timber and then leave the land without doing anything further. In many cases, continuing tax liabilities prompted the owner of cut-over land to simply abandon it so that ownership reverted to the government, sometimes local and at other times state, which took responsibility for its future use. At times the timbering process left considerable waste wood on the ground, leading to fires, and it became a governmental responsibility to fight those fires.[1]

Most of the time, abandoned lands meant the loss of tax revenue to local governments. Beyond that issue, the remnants of "timber mining" by private enterprise left a number of other problems that public bodies had to address. This role of government was a marked reminder of the ultimate public context within which private enterprise carried out its activities. Abandoned lands one way or another became a public responsibility, so governments sought ways to transform that liability and those lands into public assets.[2]

The transition that the forest reserves of the West underwent did not come easily. Many who had customarily used the public lands for their own private purposes, such as mining, timber extraction, livestock grazing, hunting game for the market, or homesteading, continued to do so. The initial occupants of the public lands (who might be termed "squatters") had long considered these activities to be a "right" that could not be denied them. As a result, much of the early history of public land management was simply an attempt to manage the land for public purposes in the face of resistance from users claiming their right to encroach on public lands. Eliminating squatter homesteads in Yosemite Valley in California, requiring stockmen to obtain permits to use forest grazing lands, prohibiting the poaching of buffalo or elk in Yellowstone National Park, and preventing timber theft in forest reserves were all phases of the attempt to establish public rather than private rights to public lands.

These competing claims of private and public rights to public lands were played out over many years and were frequently resolved in favor of the public. This controversy continues to the present day when, for example, stockmen reassert the claim that their permits to use the range are in the form of a property right rather than a limited privilege or when off-road motorized vehicle users make the even more vigorous assertion that they have a "right" to travel anywhere on public lands.

The attempt to establish the authority of public administrators such as the U.S. Forest Service to manage lands for public rather than private objectives took place not only at the level of resource use but also in the halls of Congress. Western resource users—miners, livestock owners, and lumbermen—all readily called upon their elected representatives in Congress for assistance when their claims of private rights to use public resources met resistance from the local officials of the Forest Service. At times this assistance to constituents took the form of legislation to modify forest management. The most extreme proposals were to abolish the reserves or transfer them to the states or to exclude particular lands, such as grazing lands, from the reserves and open them to sale.

In one case, petitioners successfully got Congress to pass legislation that opened the national forests to homestead settlement. The Theodore Roosevelt administration professed "antimonopoly" objectives in its resource policy and gave substance to this claim by endorsing a policy to give priority in its grazing program to the "small rancher." Congress took heed, and the Forest Homestead Act, passed on June 11, 1906, provided that lands within the national forests that were "suitable for agriculture" would be available for settlement under the homestead laws. To aid in implementing this policy, Congress went further on August 12, 1912, authorizing the Forest Service to reclassify all national forest land to determine what lands should be returned to the public domain for homesteading.[3]

The timing of efforts to "privatize" national forest lands varied over the years. A particularly contentious situation arose after the dismissal of Gifford Pinchot as chief of the Forest Service in 1910.[4] For the next decade, from 1910 to 1920, Congress continually took issue with the agency, and an observer might well have considered the Forest Service's survival to have been doubtful. The agency's own commitment and competence enabled it to survive. But the attacks, in different form depending on the circumstances, continued.

In the 1920s, as grazing became an ever more hotly contested issue because of permit fees, stocking levels, and disputes over the condition of the range, western members of Congress conducted extensive hearings in their region, hoping to arouse and organize anti–Forest Service sentiment.[5] These hearings led to unsuccessful proposals to transfer forest grazing lands to the states. Still later, in the 1930s and 1940s, due to depression and preoccupation with World War II, these congressional attacks on the agency abated somewhat. But in the 1960s and 1970s they were revived, fed by disputes with the Bureau of Land Management over grazing issues on its lands but still embroiling all the public lands.

Gradually the issues took on sharp partisan contours as the Republican Party in 1980 became a vehicle for organizing western hostility to public ownership of the public lands. President Ronald Reagan's secretary of the interior, James Watt, represented a new influence in western land affairs; he had become well known as spokesman for the Mountain States Legal Foundation, a "think tank" in Colorado that continually expounded the virtues of private enterprise and lobbied for two proposals: either sell public lands to private enterprise or transfer them to the states. Both proposals foundered, largely from opposition in the West itself.[6]

By the time the George W. Bush administration came to power, the context for the debate over public lands management had begun to change. The main source of privatization theory became the Property and Environment Research Center (PERC) at Missoula, Montana, and its leaders were early advisers to the new president.[7] But the proposals either to privatize or distribute the public lands to the states through legislation appeared to be too risky for the Bush administration; it sought to change the rules and interpret them so as to benefit private enterprise. At the same time, the western constituency for public land management had begun to change from overwhelming representation by the extractive industries to a more recently developed clientele. In this new grouping, advocates of environmental and ecological objectives had greater influence and often could partially neutralize those who were formerly dominant in western affairs. They were shaping a process leading toward greater support for public land management objectives.[8]

These more recent western attitudes toward the public lands became more manifest in the last quarter of the twentieth century. In the 1970s and on into the presidential administration of Ronald Reagan, western views toward federal ownership of the public lands were decidedly hostile and popularly known as the "Sagebrush Rebellion." These lands, so the argument went, belonged to those who occupied and used them and should either be privatized or transferred to western states. The fight over this issue was waged with considerable vehemence and media publicity.[9]

But when different versions of similar proposals appeared during the George W. Bush administration, first in the form of a mining law revision that would permit unused mining claims to be transferred to private investors for economic development of almost any kind and then an administration proposal to sell national forest land to fund local schools and roads, they both met with almost instant and widespread opposition. The discord came not only from the West but the rest of the country as well, and it revealed strong support for continued federal ownership.[10] It appeared that the earlier Sagebrush Rebellion had masked a deeper and less sensational support for public ownership and use of the national forests. This grassroots support merely awaited a crucial opportunity for expression and found it in opposing the privatizing strategies of the Bush administration.

The newer environmental and ecological interests taking shape during and after the 1970s brought to the federal public lands a more solid commitment to continued public ownership and management than the Forest Service had enjoyed in previous years. While the extractive industries had often varied in their choice of private, state, or federal ownership and management depending on how each would serve its private objectives, the newer environmental and ecological advocates were more committed to public management, and they stuck to that course. They emphasized that private industry had long demonstrated its inability to advance environmental and ecological objectives. To pursue and achieve those objectives, they argued, public management seemed essential.[11]

The Agency and the Industry

The national forests were established in a climate of distrust of the nation's private forest industry, which had been responsible for the rapid forest destruction of the nineteenth century. Their establishment was also intended to offset a looming timber shortage that would be detrimental to the nation, and public ownership of a large chunk of the nation's forest resources was intended to ameliorate some of the more disastrous problems that were foreseen. It was thus somewhat ironic that, once established, the U.S. Forest Service proceeded to sell timber from the national forests and thus continue liquidation of the nation's timber resources.

It did so because of its general philosophy, derived largely from European forest management practices, that the existing "old" and "virgin" forests in its care were no longer productive, had outlived their usefulness, and should be replaced with forests having an annual growth rate greater than that of the ancient trees; they could be made so by continuous cultivation and cropping.

While wood from liquidating the old-growth forests would add to the nation's timber supplies, the problem of a timber shortage eventually would be resolved by the contributions from this new so-called regulated forest, which would be cared for from initial planting to harvest as fully as a crop of wheat or corn. The atmosphere of the new age represented by the U.S. Forest Service was thus one of optimism about the nation's wood supply, which the agency would foster. By example, it would persuade the private forest industry to do likewise.[12]

Gifford Pinchot, first chief of the U.S. Forest Service, was skeptical about the ability of private industry to adopt such enlightened practices. Economic returns from "mining" existing trees precluded much interest in investing in a new crop; it was much more advantageous for timber businesses simply to abandon the cut-over land, returning it to the responsibility of local and state governments. Opportunities for investors were much greater in other kinds of production rather than in forest land. As a result, Pinchot did not believe that the national forests could become a model for the private forest industry; it simply was not primed to benefit from the example of the national forests.[13] Some form of public regulation of the private industry was required. The key problem was the production and cultivation of a new crop, the central focus of "sustained-yield" management. Tending a new crop was always the other side of the coin of timber harvest, and the industry was in no mood to do this voluntarily. After he left the Forest Service, Pinchot seemed to spend as much time advocating federal regulation of the private timber industry as he spent with Forest Service affairs themselves.[14]

Successive chiefs of the Forest Service after Pinchot, except perhaps William Greeley, the third chief, took up the same mission. They predicted that a "timber famine" would follow simply because private industry continued to "cut and run" and failed to stop and invest in new trees. Only public regulation could prevent such a disaster. Chief after chief proposed regulation, with Pinchot continuing the chorus from outside the agency.[15] A few bills were introduced in Congress to accomplish that objective, but they received little attention. Because of firm opposition from the industry, such proposals were dead on arrival.

Yet the issue continued to shape public debate and generated a deep mutual distrust between the agency and the industry until well after World War II.[16] It was only in 1952 that the new chief, Richard McArdle, effectively halted the campaign simply by ignoring it as an issue worthy of mention in his first annual report. Only then did the issue's disruptive role in agency-industry relationships disappear.[17] This three-decade episode was a distinctive phase of an internal battle over national forest affairs and a backdrop to the close relationships that developed between agency and industry because of wartime emergencies and that continued in succeeding decades.

One major development in the management of the nation's forests arose from solidifying the role of the states in forest affairs.[18] One step in the debate over public regulation of private forests was to admit the seriousness of the public versus private management issue but to propose regulation by the states rather than the federal government. Pinchot had insisted on federal regulation, but many who were interested in forest affairs chose the state regulation route. The result was that some states began developing forest regulatory bureaucracies in a more favorable political climate, which was considered a means of reducing the intensity of the federal versus private debate.

The major development that arose from the role of the states was a patrol-and-suppression fire management policy carried out through a federal grant-in-aid program called for in the Weeks Act of 1911 and expanded in the Clarke-McNary Act of 1924.[19] Tentative steps toward a fire program had begun to evolve in the states already, and private industry was amenable to some action simply because the uncertainty of fire as a potential threat to those who owned logging rights or forest property rendered their activities somewhat speculative and their investments shaky. Thus, it was from this common interest in more effective fire programs that there arose a more cooperative relationship between public and private forest advocates. This constituted a stage in which relationships between states and private forest industry became increasingly close, much more so than federal public-private relationships.

World War II witnessed a growing rapprochement between the Forest Service and the timber industry based largely on their common efforts to supply wood for the war effort.[20] Close cooperation continued in the immediate postwar years in a joint effort to supply the rising demand for housing construction, long pent up by depression and war. Amid this common effort to increase wood production, the agency not only dropped its criticism of the industry and demands for regulation but also made major changes in its method of calculating the allowable wood harvest on its own lands. It did so in such a way as to increase the attractiveness of its supplies to industry purchasers. The key to these changes lay in the agency's method of predicting future wood production. By inserting into its calculations favorable estimates of the effect of intensive "inputs" such as superior seed, fertilizers, and herbicides to suppress competing vegetation and treatments to control pests and diseases, the agency could "predict" larger wood harvests. It was then possible to increase the current allowable cut in line with these predictions. The entire process came to be called the "allowable cut effect" or the "earned harvest effect," and it provided a rationale for justifying increased harvests, much as the industry, which had depleted most of its own supply, demanded.[21]

This calculation method also led to a rhetorical shift from the longstanding goal of "sustained-yield" production to the new goal of "maximum production." It also gave rise to pressures within the agency to modify other statistical calculations and thereby hide its commitment to ever more intensive cutting. A further effect was that the new calculation method tended to solidify the agency's commitment to wood production as its dominant objective at the same time much of the public and the scientific world were asking it to conserve resources, in line with newer environmental and ecological objectives.

By the beginning of the twenty-first century, the nation's forest reserves had come full circle. In 1891, they were thought of as a response to the depletion of the nation's forests and the impending "timber famine."[22] These forests were to be saved from disposal to private parties so that they would forestall the predicted shortage of wood and thus solve a national problem that private industry, in its eagerness for immediate profits, could not solve. The national forests, moreover, were to serve as a model for long-term "sustained-yield" timber management that industry could emulate.

Over time, however, these roles tended to change. Important sectors of the wood products industry, faced with shortages of "virgin" timber, began to be interested in growing new forests. Amid the prevailing markets for building lumber, this effort made little headway, but once pulp and paper began to dominate, industry's perspective changed. The shorter growing cycle for pulpwood and the expanding markets for it made investments in permanent wood production plantations more economically feasible. A new wood products industry based on pulp production began to develop in the nation's Southeast. With these changes, the national forests contributed no more than a fraction of the nation's wood supply, and their role in providing the nation's goods and services shifted markedly toward environmental and ecological objectives, public benefits that private industry was singularly unable to supply. By the year 2000, the national forests were playing a new role in the national life of the country.[23]

The twenty-first century brought about an entirely new and profound change in the forest industry when, in the search for more investment returns, many forest landowners came to the conclusion that their lands were more valuable for development than for long-range management for wood production. This new perspective seemed to focus on two venues. Larger timber landowners began to analyze and divide their holdings, with the objective of selling the more attractive of their sites to individuals who wanted to build homes, while owners of smaller tracts near cities found considerable financial advantage in "liquidating" their woodlots—selling them to developers who were fostering the "sprawl" taking place around urban areas.

As the largest forest landowner in the country, the Plum Creek Timber Company, advanced eastward from its home country in the northern Rockies to establish roots in New England, its development plans aroused considerable opposition from citizens. New Englanders were appalled at the prospect of losing "traditional uses" of their forest lands.[24] The Forest Service took considerable interest in the use of conservation easements to protect land from development for continued wood production around cities, championing the use of federal funds in the Forest Legacy Program for easement purchases. In 2002, the chief of the Forest Service, Dale Bosworth, identified sprawl as one of the four main threats facing the agency in the twenty-first century.[25] The threat of sprawl, especially in the nation's Southeast, tended to establish a new partnership between the industry and the agency.

Transition in Agency Publics

The broader public context within which the Forest Service carried out its management mission differed vastly at the beginning of the twenty-first century from that of its earlier years. Activities such as recreation, which was a barely noticeable part of the world of forestry in 1891, had become of vast importance or even dominant more than a century later. At the time of its founding, the agency rejected the notion of national forests as venues for home, work, and play or as wildlife habitat, considering those objectives to be inappropriate. Public interest in those objectives became even more active during the twentieth century. Watersheds, which ranked high on the agency's initial agenda in the 1890s, were a recognized but less urgent issue throughout the next century.[26]

The agency's initial concentration on its clientele of timber producers and ranchers engaged in extractive grazing prompted it to downplay and even reject the importance of other potential objectives, but over time it was forced either by circumstance or by law to give other forest uses greater recognition. Finally, toward the end of the twentieth century, a new public context arose, emphasizing the importance of the environmental and ecological resources of the national forests and challenging all users to focus on resource sustainability amid pressures from increasing human use.

In 1891, at the legislative birth of the national forest system, the reserves were of importance to many sectors of the public. Some Americans looked upon them as amenities with potential for national parks.[27] For others they promised an important source of wood as other sources of future supply continued to decline; some viewed them as habitat for game that needed protection so that depleted populations might be restored.[28] Still others thought of the forests as watersheds to protect water supplies.[29] But as the reserves were transformed into national forests, this range of objectives was narrowed considerably to a primary focus on grazing and timber, the two most visible extractive activities. These issues dominated the agency's objectives for a number of decades. How this change came about is a major element in the story of the American public's relationship with the national forests.

Gifford Pinchot was the architect of the initial public role of the national forests and the U.S. Forest Service. He deliberately rejected the notion of forests as wildlife refuges or as public amenities.[30] He also gave little attention to watershed objectives and subordinated them to the more important grazing and wood production programs. Amid these various public objectives expressed in the first decade and a half of the forest reserves, Pinchot brought economic development to the forefront of forest management, making economic activities such as grazing and timber production integral parts of the Theodore Roosevelt administration's economic conservation program. Thus, the agenda of the Governors' Conservation Conference of 1908, which Pinchot had an influential hand in drafting, did not give much attention to wildlife or aesthetics and focused on water resource development rather than watershed management.[31]

Decisions that considerably narrowed the focus of national forest management gave rise to separate streams of conservation action. Rejected by the development policy makers in the Theodore Roosevelt administration, those parties favoring a focus on wildlife or "amenity objectives" looked elsewhere for support. They sought special legislative and executive authorization for programs that were then established without connection to the national forests. These efforts to enhance both amenity and wildlife objectives resulted in the development of the National Park Service on the one hand and state and national fish and wildlife programs on the other.

From 1920 to 1960, some of these proponents of wildlife and aesthetic issues who had earlier been excluded from the national forest agenda sought once again to be part of national forest affairs. Wildlife advocates broadened their constituency from the wealthy big-game hunters who were represented by the Boone and Crockett Club to local and regional sportsmen's clubs with a wider membership, and by the 1930s, they had begun to achieve greater influence in public affairs.[32] The automobile gave hunters more ready access to the national forests. The Forest Service responded by establishing the Division of Wildlife in 1937.

In similar fashion, automobile camping brought an ever-larger number of people to the national forests. At the same time, the growing demand for the Forest Service to establish wilderness areas for aesthetic enjoyment, an objective that Pinchot had rejected, led in 1964 to legislative authorization for such areas. By the 1950s, this new group of users demanded to be recognized in the agency's authorizing statute. That demand led to the Multiple-Use and Sustained-Yield Act of 1960, which put a de jure stamp of legal approval on users and uses that were already playing a role in agency affairs even though they were not yet legally recognized. The agency's constituency in 1960 was thus vastly different from that of 1905.

In the second half of the twentieth century, that constituency changed even further, and with a twist that presaged a new direction in agency affairs with implications far beyond the simple addition of new uses and new users to old. In the decades after World War II, a public of citizens and scientists became more interested in the national forests as reservoirs of a much wider range of resources, evolving from game to nongame wildlife, then to a wider range of plants and animals, and finally developing into a focus on biodiversity as the way to think about wild forest resources.[33]

This emphasis was brought to bear on forest objectives especially through participation in the planning process established in the National Forest Management Act of 1976. It involved an emphasis not so much on user opportunities as on the detrimental effect of increasing numbers of users on forest "wild resources"; it required that uses and users be curbed in order to protect and enhance the forests' ecological health. Ecological forestry also received more recognition from the Bill Clinton administration and became a significant element in the work of the Science Advisory Committee, appointed by that administration to revise regulations to administer the 1976 act, including its proposal to make "ecological sustainability" a central objective in the administration of the national forests.[34]

These evolving constituencies of the Forest Service brought to agency affairs an increasingly well informed public and independent sources of knowledge that could be brought to bear on a wide range of forest management issues. Many citizen organizations were formed to focus on a particular national forest or forests in a specific region. Such organizations had considerable personal knowledge and systematically organized information at their disposal and could track affairs on particular forests and respond quickly to agency proposals.

In some cases, individuals with a special interest in a particular aspect of forest affairs such as endangered species became a source of expert knowledge on which citizen groups could draw. Often these individuals were as well informed as agency staff and therefore could meet the agency on even terms in the world of public debate and decision making. Yet the Forest Service seemed to be almost oblivious to this significant public development. Instead, the agency attributed public criticism less to informed citizens and more to the dramatic and media-fostered images of tree sitters and street demonstrators.[35]

Forest Science and Forest Professionals

The scientific and professional context of forest affairs changed considerably over the years of the Forest Service's existence, at times reflecting the agency's need to deal with new responsibilities and, in the late twentieth century, mirroring the interest of scientists in the newly observed ecological conditions of the forests. The national forests were originally designated during a period in which the "scientific culture" purported to provide a superior way of dealing with public issues. With the new scientific culture, public administrators were selected for their professional skills rather than for their patronage of a particular elected official. This early pretension to forest science involved not just a method of hiring staff and making decisions but also an actual science: silviculture, or the cultivating of forest trees. It was the subject matter of science, not the method, which changed in the agency over the course of its first century, changes which reflected changes in the agency's management directions.

In the early days of the forestry profession, one spoke of "forest science" without really specifying that "foresters" were not only expounding a method of acquiring knowledge but also identifying the subject matter—wood production—about which information was to be obtained through "science."³⁶ Central to the subject matter of wood production was a sharp revision of the required knowledge, shifting from a broad-based botany to a specialized segment of botany representing trees and shrubs, a subject known as dendrology. Forestry students were advised not to learn the wide-ranging botany in which the pioneers of modern forestry had been trained but to focus on a more narrow range of species, primarily those of commercial value. By 1950, the most widely used dendrology textbook, one written by William Harlow and Ellwood Harrar, advised the reader that "it is felt that students of forestry should first know well the commercial species of North America."³⁷

Young forest professionals were immersed in this selective subject matter so much so that over the years it was virtually impossible for them to develop as thorough a "scientific" approach to other forest resources. The Society of American Foresters (SAF), the professional body of forest specialists, had a similar difficulty in taking a broader view of forest biological resources.

One of the SAF's members, Leon Minckler, former director of the branch of the Northeastern Forest Experiment Station located in the Shawnee National Forest, attempted in the 1970s to broaden the society's perspective beyond its limited focus on wood production. He advocated what he called "ecological forestry" and spearheaded a petition drive. His petition, signed by 151 professional foresters, asked the society to establish a special work group (a common practice in the society) on ecological forestry. The petition was denied with the argument that, since all forestry was ecological, there was no need for such a working group. Rebuffed, Minckler then drew up a series of pamphlets on ecological forestry for different forest regions and published them with the National Parks and Conservation Association.³⁸

This sharply delineated focus for forest science was closely associated with a similarly sharply limited concept of management. Forest Service staff and other professionals continually argued that such matters as wildlife, recreation, and amenities called for a "nonmanagement" approach, that is, one did not "manage" wildlife, recreation, or amenities.³⁹ These areas were often identified as areas of "special uses" in contrast with the larger "general forest" that was devoted primarily to wood production.

To implement its scientific objectives, the agency established an elaborate system of research in its forest experiment stations, which were given statutory authorization in the McNary-McSweeney Act of 1929. These continued to provide scientific information about "production forestry," a term frequently used to refer to the science of wood production.[40]

At the same time, forest education forged ahead rapidly. By World War I, twenty-four forestry schools had been established and twenty of them were still in operation. Their curriculum was almost exclusively devoted to wood production.[41] The desired course of study was outlined by a group headed by Henry Graves, Pinchot's successor as chief of the Forest Service and first dean of the Yale School of Forestry. Nearly half of the proposed curriculum was devoted to silvics, management, and mensuration; more than a quarter to forest utilization and products; the rest to protecting forests from fire, insects, and disease.[42]

As the agency was forced by circumstances to accept new forest objectives, it seemed to convey almost implicitly the need to "manage."[43] For example, the agency embraced the idea that recreational users had to be "managed" by isolating them from the general forest and requiring that their camping be confined to selected areas. At the same time, ongoing conflicts among recreational users, for example, between hikers and motorized vehicle users, required the Forest Service to have some idea as to "what those users wanted," and thus the field of "sociological" research arose to provide information as to how one group of users could be accommodated without arousing the hostility of another. The agency promoted research of this kind but called it "recreation" rather than "forestry" research. This nomenclature perpetuated the notion that "forest science" referred only to wood production and not the entire range of forest-related activities and circumstances.

Wildlife highlighted quite different forest resources because the knowledge base regarding the overall ecology of forests and how that knowledge should be organized for systematic understanding (i.e., research) differed markedly from the information and strategies inherent in traditional forest science. The species of interest in the larger area of forest ecology were not just the few dozen plants containing commercially valuable wood fiber but some fifteen thousand additional species of plants and animals in the eastern hardwood forests alone. Each of those species, if seriously managed, could require the same degree of scientific and professional attention as did wood-producing species. Moreover, wildlife management required close attention both to habitat (which was the unit of observation and study in wildlife science) and to the relationship between species and their environment. The scope of habitat as a unit of study stood in marked contrast to the scope of the "stand," which was the unit of observation and management in wood production. Thus, an expanded view of what a forest constituted required a fundamentally different orientation than the traditional notion provided by forest science.[44]

In the last third of the twentieth century, as the notion of wildlife evolved from a narrow view of game animals to a wider view encompassing nongame species and then biodiversity, encompassing the full complex of plants and animals in the forest ecosystem, the world of forestry acquired an entirely new dimension. This new aspect included the comprehensive identification of forest species, their habitats and relationships, changes in population levels over the years, and the impact of human "disturbances" on forest ecological conditions.[45]

There was thus a new definition of what a forest was, and this new paradigm generated a wide range of new scientific work with an ecological dimension. These major changes of focus in Forest Service activity came about as a result of mandates in the National Environmental Policy Act of 1969. That act required that the Forest Service obtain comprehensive, "searching," and "interdisciplinary" analyses of the environmental consequences of its policies and projects. To complete these analyses, the Forest Service had to call upon specialists whom the agency did not traditionally employ, and this brought a new group of natural scientists and new facets of forest science to its work. The new staff included ecologists, zoologists, botanists, and those specializing in nongame as well as game wildlife. These experts were found not in forestry schools but in other academic departments, which led to new and more extensive types of associations between specialists in the Forest Service and their counterparts in academic institutions and other federal resource agencies. These changes tended to fracture a professional culture that was formerly dominated by silviculturalists and engineers. As these new specialists gradually filtered into the Forest Service, they constituted a group quite different from the agency's traditional professionals, and they came to be known within the agency as the "ologists."[46]

On occasion, the different perspectives of "production" scientists and ecological scientists were rather sharply displayed. For example, in the spring of 2002, two groups of scientists engaged in an exchange with President George W. Bush. On April 16, a letter to the president signed by 220 ecological scientists questioned the wisdom of commercial timber harvesting in the national forests. Describing themselves as "conservation-minded scientists with many years of experience in biological sciences and ecology," they stated that thirteen thousand plant and animal species lived in the national forests. The scientists argued that "it is now widely recognized that commercial logging has damaged ecosystem health, clean water, and recreational opportunities." Most of the signers were associated with a wide range of academic departments, not one of which was a forestry school.[47]

Two weeks later the president received another letter, this one signed not by scientists but by the presidents of the Society of American Foresters, the National Association of Forestry Schools and Colleges, and the National Association of State Foresters. This letter asserted the importance of timber harvesting for the national forests, implied that the authors of the first letter did not represent the scientific forest community, and asserted that they were the more valid voices for forest science. An accompanying report by the Society of American Foresters advised the president to ignore the ecological scientists "in favor of forest management proposals supported by credible scientific evidence."[48]

Management Capabilities

The capacity of the Forest Service to manage the resources under its charge grew steadily over the course of its first century, but management demands grew at an even greater pace. The agency's growing clientele, with needs and demands that required it to oversee more than just wood production, taxed its managerial capacity severely at times. The self-image of the agency was that it was a "can-do" organization that could tackle whatever task it faced, but as the century wore on and the "multiple-use" philosophy was followed by "environmental" and "ecological" forestry, the ability of the Forest Service to do a satisfactory job was sometimes woefully limited.

The overriding agency concern from the start was fire—how to suppress forest fires once started and how to prevent them in the first place.[49] Fires had long been the bane of forest communities and were sharply ingrained in the public mind by spectacular ones in the Great Lakes region in the last third of the nineteenth century. By the time the national forests became the focal point of firefighting and fire prevention research, forest firefighting and fire suppression had already been around for several decades. The Forest Service had a labor force at hand to act as firefighters. No matter what one's job was in the national forests, employees were expected to drop their assigned duties and join the fire brigades at a moment's notice.[50]

After several years of experience with firefighting, the Forest Service formulated an agency-wide "ten o'clock" policy—the goal that every fire detected one day would be contained by ten o'clock the following morning. This ambitious objective called for increasing management resources, and over the years those resources evolved steadily. The Civilian Conservation Corps (CCC) of the 1930s provided a new and expanded labor force for labor-intensive management tasks, among which firefighting was the most demanding. Specialized firefighting personnel were organized and trained; more effective methods of detecting fires were developed; speedier communication systems were applied; "smoke-jumping," involving firefighters parachuting to fires in remote areas, was instituted; and aerial drops of fire retardants became routine. Firefighting ate into not only the agency's management responsibilities but also its budget.[51]

Fire prevention soon took its role alongside fire suppression and became an increasing part of Forest Service fire management. At first, prevention activities focused on the human role in starting fires, with publicity campaigns urging the new hordes of forest visitors to exercise greater care. The agency sponsored media campaigns, the most extensive being the one that featured Smokey Bear with the admonition that "only you can prevent forest fires." Most fires were "natural," caused by lightning, and fire prevention increasingly came to emphasize the fact that fires were made more severe by the forest floor's accumulation of old vegetation, a potent fuel source. Burning slash, which consists of cut-off tree limbs, unwanted brush, and small trees, to reduce its role in turning small fires into larger ones was a practice that Pinchot required of any timber company that wanted a contract to harvest in national forests.[52]

Both fire suppression and fire prevention programs became increasingly more complex toward the end of the twentieth century because development had pushed into forest areas.[53] These areas were most attractive to urbanites who sought either a vacation home or even a permanent home in the "woods." This urban fringe in the inhabited forest zone surrounding cities stimulated more human-caused fires, led to demands by forest home owners for more fire protection, and, in turn, led to increasing pressure from rural firefighting units, insurance companies, and the Forest Service itself for residents to take more action to insulate themselves from potential fire damage. People in the residential zone played an important role in federal support for the fire program by insisting that more funds be spent in their location rather than in the deep forest where there was little or no human habitation.

Forest uses by people, authorized or unauthorized, required quite a different type of management: enforcement. Some uses could and were gradually brought under control through use permits, which was a matter of drawing up rules and then having sufficient personnel to supervise their enforcement. But personnel were also required to deal with unauthorized users—a situation that called for negotiating skills, which field personnel were not likely to possess.

From the very start of the agency in 1905, it was clear that these supervisory tasks were massive. The territory to be supervised was extensive, and many an activity could be carried on by those with imagination, experience, and cunning in the many hours and days between visits by staff. Such circumstances applied to a number of illicit activities outlined in the agency's "use book," which served as a bible of directions to personnel in the field.[54]

As time went on, even more people sought to use the national forests for a variety of purposes, some illegal, such as growing marijuana, and management challenges became more extensive over the years.

As public access to the forests increased with the boom in road construction and automobile travel, the agency sought to bring these new users under control by designating where camping could take place. As more users began to arrive with trailers and other recreational vehicles, the agency had to update their campgrounds with larger spaces and water and waste facilities.[55] Designated campsites were established in the backcountry, and the agency developed rescue facilities for injured hikers. So many people began visiting these wilderness areas that the Forest Service established a permit system to limit the number of users.

Later on, forest users became even more mobile with the boom in off-road vehicles (ORVs). These machines presented an even more difficult problem, especially because they made their own trails through the forests. For many years, however, the agency took only limited action to control them. Agency staff debated whether they should be confined to designated trails and prohibited elsewhere, in what was called a "signed-open" policy, or prohibited in designated areas and permitted elsewhere in a "signed-closed" policy. Much of this problem was left to each forest supervisor to deal with, resulting in policy inconsistency across the national forests.[56] In 2003, agency chief Dale Bosworth sought to require each national forest to adopt a control policy. Whether or not this would bring ORV use in national forests under control remained to be seen.[57]

These public-driven multiple uses of the national forests placed the most significant pressure on the Forest Service's management capabilities. The diverse array of recreation enthusiasts often got into each other's way, and the agency spent considerable time finding out what each group wanted and how they could be kept from interfering with each other. One way was to provide separate trails for hiking, horseback riding, and motorbiking. The agency's ability to handle these specialized management tasks varied. For the traditional activities of wood production and grazing, the Forest Service had gained considerable experience, developed needed technical information and management skills, and approached its tasks with considerable confidence. Initially lacking that sort of expertise in managing recreation, the agency had to hire specialized personnel in almost every national forest. These new employees were not thought of as "foresters," however, and the agency even placed them in a separate personnel category.

Environmental and ecological management objectives after 1970 challenged the Forest Service to develop new ways of managing wildlife and watershed resources. It was ill prepared to establish firm management directions for either area because both objectives had been neglected throughout much of the agency's history. Hydrologists assigned to management positions rather than scientific research were not appointed until after World War II, and the agency's first separate wildlife administration was in 1937.[58] The agency's usual practice had not been to provide continuous management for those areas but to devote only enough resources and staff to ensure that the agency complied with the law. Both areas were shaped heavily by evolving scientific knowledge, which could be applied to achieve compliance or enhance management, and the agency chose to confine itself primarily to compliance. It looked upon its role in the evolving science more as a task of keeping informed so as to avoid legal action against it rather than of developing information and professional capabilities for more full-scale management objectives.[59]

These new ecological and hydrological objectives in forest management presented the Forest Service with yet another massive challenge. The agency was continually called upon to take various actions even though its knowledge of these areas was only partially developed. In response to this challenge, there arose the notion of "adaptive management"—the idea that management would be instituted but that it would continuously adapt its practices to take the growing fund of knowledge into account.[60] Forest users, in turn, and especially the wood-using industries, sought protection against adaptive policy changes on the ground that economic investments required greater stability and protection from such "uncertainties."[61] With ecological forest knowledge continuously advancing, new skills were called for, especially the ability to work effectively with less than complete knowledge. This skill was at a premium among agency personnel, who were accustomed to a tradition of much firmer management practices.

These century-long developments in Forest Service affairs chart the major patterns of changes in the agency. The three chapters that follow offer a basis for historical understanding that provides more insight into the processes of change that are specific to time and place. In doing so, they stress the fact that, at any given time, the agency functioned in the present and faced the future with an inherited legacy of institutional practice and responsibility. By 2005, the agency had operated under a series of three legacies (corresponding to the three chapters that follow), and each legacy was the result of a distinct period of institutional experience and set of relationships between the public and the agency.

2 THE SILVICULTURAL IMPERATIVE, 1891–1920

TRADITIONAL accounts of the early history of the U.S. Forest Service have emphasized its legislative and administrative development in a legal context.[1] The customary historical benchmarks have long been the reserve authorization of 1891, the Forest Management Act of 1897, and the Transfer Act of 1905, which lodged the management agency permanently in the Department of Agriculture. Other critical events in Forest Service history are court decisions such as *Light v. United States* and *United States v. Grimaud*, both decided in 1911 by the Supreme Court, which upheld not only the agency's grazing permit decision-making authority but also its rule-making authority under the 1897 act.

Within this overall framework, historians recount the steady evolution of the agency's management system as shaped by Gifford Pinchot, the nation's most influential forester at the time; the emphasis on expertise for agency staff; the development of effective working relationships between the central office in Washington and the field; and the development of an employee esprit de corps that was unusual for government agencies at that time.[2]

Gifford Pinchot, however, did far more than foster the development of an effective administrative system; he chose to shape the specific direction of the agency's program. In doing so, he greatly narrowed the agency's forest management objectives, going from the wider context of the 1890s debate over the reserves to the narrower context of a comprehensive program for the national forests, which evolved after 1905.

In 1891 and in the decade that followed, various management objectives were expressed, and they revealed different views of the purposes of the emerging forest reserves. Some looked upon the forests as a habitat for wildlife and, in the midst of declining wildlife populations, thought of them as refuges where animals could be protected and their numbers increased.[3] Others thought of the forests as cover to protect their sources of water for irrigation and urban and industrial use and as stream habitats for fish.[4] Still others thought of them as sources of timber to offset an impending "timber famine," while some looked to the forests for their natural beauty and as lands to be designated as national parks.[5]

In the late 1880s and early 1890s, those interested in forest affairs formed their ideas based on the context of the Adirondack region of upstate New York. This region was the center of the nation's wood production industry in the mid-1800s. It had been heavily cut over, and by the 1880s, it was the model of just what should be avoided because it was fast becoming, according to an Adirondack study commission headed by Harvard professor Charles Sargent, an "unproductive and dangerous desert."[6]

In 1885, the New York state legislature reserved from sale all tax-reverted lands and declared them to be a forest reserve "to be kept as wild forest land."[7] By 1892, these lands had been designated the Adirondack Park, the territory of which was bounded by the "Blue Line," within which about 20 percent of the 2.8 million acres was owned by the state, which anticipated acquiring even more land.

How would this land, the nation's first major forest reserve, be managed? To the New York legislature, the reserve was primarily a watershed to protect the sources of the state's rivers and a "pleasuring ground," so the state had only minimal interest in the timber supply. Thus, the legislation stated that the new park would "be forever reserved, maintained, and cared for as ground open for the free use of all the people for their health and pleasure, and as forest lands necessary to the preservation of the headwaters of the chief rivers of the State, and a future timber supply."[8] The fact that the Blue Line was drawn to encompass the headwaters of Adirondack rivers made clear that a primary objective of the park was watershed protection.

Both Bernhard Fernow, the head of the federal Bureau of Forestry in the last years of the nineteenth century, and Gifford Pinchot, who succeeded him in 1898, followed the developments in the Adirondacks with great interest and intense opposition, almost contempt. They recognized that policies governing the use of Adirondack Park focused on watershed protection and aesthetics and seemed not even to acknowledge their version of scientific forest management, which placed primary emphasis on the economic value of wood production.[9] Forest land for primary use as watershed protection or for aesthetic enjoyment was not on Pinchot's agenda, so to him, the Adirondacks represented what forest management should not be. To the entire emerging profession of forestry, in fact, the Adirondacks remained not only "forever wild" but "forever castigated," as an example of what "true" foresters should reject.

The Forest Management Act of 1897 narrowed the range of possible forest objectives to just two: wood production and watershed protection. At the same time, the reserve managers effectively added livestock grazing to the accepted list of objectives and rejected two others, wildlife refuges and forest aesthetics, while giving little attention to watershed protection. This agency perspective, implemented initially by Pinchot, set the direction for agency management for many years to come.

The early years of the Forest Service tell the story of how Pinchot narrowed the wide range of possible forest objectives to the limited program he favored.

Grazing Management

The national forests were established ostensibly to respond to the nation's potential "timber famine," but until after World War I, the program of the

U.S. Forest Service was dominated by its role in the western livestock industry.[10] From a statutory point of view, this situation was more than remarkable since the Forest Management Act of 1897 provided no legislative authorization for grazing, and in the debates leading up to that act, grazing was, in fact, considered to be detrimental to national forest policy.[11]

Foresters believed that livestock, especially sheep, grazing in the forest headwaters caused soil erosion and denuded the landscape, thus threatening the watershed and water users (such as irrigation farmers and urban residents) and also the reservoirs, in which they sought to store water for their own use.[12] This was no minor controversy but one that had continued for years prior to the enactment of the 1897 law and the establishment of the Forest Service in 1905. So despite the absence of grazing authorization in the 1897 act, the years running up to it and the transfer of the reserves to the Department of Agriculture were heavily influenced and even dominated by grazing issues. From the early years of the national forests, the agency was thrust into the midst of long-standing western grazing controversies.

The most important of these controversies involved the question of who was to use the range. The forage was limited and the user claimants many. The most volatile conflict was between sheep ranchers and cattle ranchers. Over the years the attitude that ranchers had a "right" to use all public lands for grazing had resulted in a free-for-all that frequently erupted into violence. This situation involved the national forests, which contained a considerable amount of grazing land, as well as the western range as a whole. Although the Forest Service was primarily interested in wood production, it had inherited much range land and with it, many range controversies.

Owners of livestock were far more numerous users of the reserves than were commercial users of wood, and as far as western leaders were concerned, the success of the reserves and national forests would rise and fall according to how the grazing lands were managed. The continued existence of the national forests depended on the ability of the Forest Service to deal with the grazing problem, so it found itself preoccupied with the issue until well after World War I. In fact, it has been argued with good reason that the personnel who worked effectively with livestock owners were crucial in turning potential agency disaster into survival.

Pinchot well knew the circumstances within which the new forest administration had to work. He had made several trips to the West and learned about the problem firsthand. He came to know one of the leading cattlemen of the Southwest, Albert Potter of Arizona, who argued that allocating rangeland to individual users under a permit system was essential to bringing some degree of order to the range. This proposal seemed to favor the less mobile cattle industry, which combined higher elevation summer grazing with lower elevation ranches that produced forage for winter feeding. Pinchot adopted Potter's ideas and brought him into the Bureau of Forestry as an assistant chief forester in charge of grazing.[13]

These moves tended to forge working relationships between foresters and the range users of the forest grazing lands. They also led to establishment of the permit system in the new national forests, produced significant support for the reserves among the most volatile of the agency's potential clientele, and, it seems safe to say, helped to protect the agency from demands that the reserves be abolished. Pinchot's most controversial move with the stockmen came when he instituted a fee for the permit, an action that led to legal proceedings to challenge the permit-and-fee system. This fight went all the way to the Supreme Court, which upheld the agency's authority to carry out the program.[14]

Throughout the early years of the Forest Service, until well after World War I, supervision and management of grazing was the agency's dominant activity.[15] This is not to say that the initial stages of wood production management were not under way; agency staff began to survey and inventory the forest's wood resources throughout the West.[16] But the grazing problem demanded more immediate attention, often more than the national office desired even into the 1920s. Thus, David Clary reports, "Ira Mason [director of the agency's timber program after World War I] was most displeased to learn that forest supervisors and district rangers spent most of their time working with those who held grazing permits, leaving timber work to inadequately trained or supervised junior staff."[17]

For a number of years, grazing continued to be more important economically than timber. Filibert Roth, in charge of the reserves in the U.S. General Land Office prior to the transfer in 1905, reported that "grazing in the forest reserves today is of greater importance, financially, to the people of the respective districts than is the timber business."[18] And Pinchot himself reported in 1905 that "at present only in rare instances does the value of the timber annually taken from the forest reserves approach the value of the forage yearly consumed by grazing animals. Perhaps only in a single instance—that of the Black Hills—is timber cutting distinctly the more important industry."[19]

Receipts from the fees that the Forest Service established provided more income than the sale of timber. Even by World War I, income from the two sources, range and timber, was roughly equal. This income was of great importance to the agency since Pinchot had argued in support of the transfer in 1905 that he could bring a profit to the government, and for many years the agency could not make good on that claim by selling timber alone.

The dominance of grazing in national forest management began to wane with the resignation of Albert Potter in 1920.[20] The issues that grazing had brought to the agency did not diminish, however; instead, they entered a new phase that pitted the desires of range specialists within the agency to improve the quality of the forage against the desires of range users for more influence and control over their activities.

This initial preoccupation of the Forest Service with grazing, it should be noted again, involved an activity that at the time had no statutory authorization under the Forest Management Act of 1897, which limited reserve management objectives to wood production and watershed protection. Grazing was not authorized formally in national forest management until the Multiple-Use and Sustained-Yield Act of 1960, legislation for which the livestock industry had long pressed. At the same time, watershed management, which was authorized in 1897 and for which grazing was thought to be detrimental, received little management attention in subsequent years as the agency focused on grazing and wood production. The role of these two objectives in agency affairs was marked testimony that the early history of the agency was shaped not so much by legal mandates as by the economic and social circumstances within which the agency made its choices.

Watershed Protection

For many years watershed protection had been intimately associated with the concern about forest denudation, and this was long before the forest reserves were established. George Perkins Marsh's *Man and Nature,* published in 1864, had been the most prominent of writings about the subject. Marsh drew not only on his experience with watershed changes in the East but also on his observations in Mediterranean Europe, which he described as having experienced long-term watershed degradation. He issued dire warnings that unless action were taken, the same fate would befall America.[21]

The interest in watershed protection had many sources. Earlier in the nation's history, many a town and city had acquired watershed lands in order to protect their drinking water supplies. The decline in fish populations was often attributed partly to forest watershed destruction. Watershed protection was advocated by industrialists who depended on reservoirs for supplies to run their water-powered mills and factories. In the late nineteenth century, merchants of New York City gave strong support to the move to protect the Adirondacks on the grounds that the canal and river navigation on which their commerce depended required it. The state of Pennsylvania, drawing heavily on the New York action, established its state forest program in 1898 with the requirement that, for watershed protection purposes, the state would purchase lands in the headwaters of its three main rivers: the Delaware, the Susquehanna, and the Allegheny.[22]

The notion that forests could affect streamflow was publicly acknowledged in the legal language of the Forest Management Act of 1897 with the provision that one objective of forest management was to create "favorable conditions of streamflow."[23] The connection between forests and streamflow was somewhat controversial, however, and the Forest Service began its work in the midst of that controversy. Leaders of the timber industry denied that logging in headwater areas could reduce streamflow, but the emerging forest professionals continued to adhere to the firm statement that the connection was clear.

Bernhard Fernow was one of these forest professionals, and he frequently pointed to long-term research in Germany that demonstrated the connection. In later years, however, he admitted that they had overstated the case and defended the theory of the relationship between forest cover and streamflow because otherwise "we could not have committed the federal government to purchase lands for National Forests."[24]

Interest in the role of the western national forests in watershed protection arose from a different circumstance: the tendency of harvested forest lands and heavily grazed headwaters to increase soil erosion, producing sedimentation in irrigation ditches and reservoirs and reducing their value for storing both irrigation water and urban water supplies. Throughout the West, a markedly water-short region, one could find people who were concerned with water supplies and who supported the forest reserves. This was especially true in southern California, where a number of forest reserves had been established at the initiative of water users. Several forests in the Southwest were established primarily as "watershed forests" to protect water supplies.[25]

A distinctive twist to watershed use made the matter highly controversial. The issue was the role of sheep grazing in denuding forested watersheds, which facilitated erosion, causing sediment to impede the flow in irrigation ditches and reducing reservoir storage capacity. Leaders of water supply/erosion interests came into sharp conflict with sheep operators, who argued that sheep played no such destructive role. Establishment of the reserves did not resolve but only continued the vigorous debate over the connection between sheep grazing and watershed protection. In its report on management objectives for the reserves, a National Academy of Sciences committee had taken the side of the water users and agreed that sheep grazing was detrimental. Their report therefore emphasized watershed protection as one of the two desired goals of reserve management and made no provision for grazing as an explicit statutory reserve objective—views that resulted in the absence of grazing as a mandated forest management objective in the Forest Management Act of 1897.[26]

The Forest Service inherited this controversy, and in doing so Gifford Pinchot sought to establish grazing as a legitimate agency objective while minimizing agency interest in watershed affairs. Watershed protection continued to be affirmed by the Forest Service, but at the same time it seemed to take a backseat in the agency's management strategies.

In the water-short West there emerged the idea that standing forests reduced water supplies by transpiring water into the air but that cutting down trees made water more abundant. The agency took this argument seriously. In 1910, near the Fremont Experiment Station, located in the Rio Grande National Forest at Wagon Wheel Gap in Colorado at the headwaters of the Rio Grande, the Forest Service established the Wagon Wheel Gap Experiment Station to study the effect of forest cover on streamflow.[27]

The agency's annual reports continued to mention ongoing studies of the effect of forest cover on streamflow. A few years of this comparative treatment led to the conclusion that the cut portion produced a slight increase in water yield but, at the same time, an eight fold increase in erosion. It seemed that the results might well have settled the issue, but the agency continued to consider it an open question and to be pulled in two directions, thus minimizing the firmness of its commitment to watershed protection.

Would the agency modify its wood harvest program to respond to the pleas of communities to protect their water supplies? The Forest Service continued to hesitate, and in order to pursue both grazing and watershed protection, it conducted experiments seeking to demonstrate that wood could be harvested and replaced with grass. These experiments were based on the supposition that the grass would produce less evaporation than would forested areas and, at the same time, would control soil erosion. The hesitancy of the Forest Service to take up a vigorous watershed protection program seemed to propel this agency objective toward a "further research" agenda rather than an action agenda.[28]

Wildlife Conservation

The decline in the population of game was as marked as the decline in their forested habitat, and consequently the move to establish reserves in forested areas for their protection elicited strong support from sportsmen.[29] This was especially true for areas that were home to bison, bear, deer, and turkey. Long since gone or rapidly disappearing in the late nineteenth century in the East, these species of wildlife had become linked in popular imagination to the western forests, where they still existed but required protection. It was the hunters interested in these game species who were most actively interested in wildlife conservation, and their primary voice came to be the Boone and Crockett Club, a group of hunters primarily from the New York City area and organized in 1884 through the initiative of Theodore Roosevelt and his friends. At the same time, a New York publication, *Forest and Stream*, under the inspired leadership of George Bird Grinnell, campaigned for the protection of forests and streams and their fish and game in both the East and the West.

This complex of interested people provided active support for Yellowstone National Park, which to them was primarily a refuge for bison and elk. Their focus was protecting those animals from poachers, and they were arguably more influential than even those interested in the forest economy among those lobbying for the vital piece of legislation that in 1891 gave the president the executive power to establish forest reserves.[30]

The first forest reserve, the Yellowstone National Timberland Reserve was, in fact, thought of as an addition to Yellowstone National Park or an expansion of the big game reserve provided by the park itself. Western forests in which big-game hunters had found opportunities to practice their sport were frequently identified as possibilities for further reserves. The Boone and Crockett Club continued to be the most significant advocate of expanding the reserves. It made no distinction between reserves intended to protect forests versus reserves designed to protect game, and as the forest reserve program evolved, the club sought to make sure that the forest reserve administrators provided for the management of game as well as trees. A 1902 bill foreshadowing the legislation that eventually transferred the forest reserves to the Department of Agriculture in 1905 in fact included a provision making them fish and game reserves as well, but this provision did not survive in subsequent versions of the legislation.

According to forest historian Henry Clepper, "Wildlife preservationists, led and supported by President Theodore Roosevelt, advocated that the forest reserves be made inviolate wildlife refuges under military protection in the manner of Yellowstone National Park."[31] But Pinchot objected to this idea and convinced Roosevelt that with hunting permitted in the forests, "game habitat and population would improve." As a result, "a certain element in the wildlife conservation movement was alienated and the Forest Service was long submitted to criticism."[32]

Thus, as Gifford Pinchot shaped the direction of the national forests, wildlife played a minimal role in his strategy. In fact, wildlife almost disappeared from the agency's agenda, an agenda maintained only by a few on-the-ground Forest Service personnel rather than by its policy makers. The notion that wildlife reserves would be an integral part of the forest reserve system evolved such that a separate set of wildlife reserves was established through a separate set of executive orders. For example, the Wichita National Forest, established by Roosevelt in 1900 as a reserve in which bison were to be protected and propagated, became by separate executive order in 1906 the Wichita National Wildlife Reserve. At times, failed legislative efforts to establish a national park that included wildlife protection objectives, such as in the Grand Canyon, led to a presidential proclamation to establish a wildlife reserve, in this case the Grand Canyon National Game Preserve to protect the kaibab deer. Historians have yet to delve in detail into the way in which the new national forest administrators rejected wildlife reserve objectives as part of their responsibilities.[33]

A good illustration of the tenuous role of wildlife objectives in forest administration is the frustration experienced by a young Forest Service employee, Aldo Leopold, who took up a post in New Mexico in District 3 of the national forests in 1908. Leopold, who had just finished a forestry course at Yale, was on his first tour of agency duty. He was enthusiastic about the new forestry business of which he was now a part. But Leopold had a wider set of interests than just wood production. So along with playing a part in the agency's new task of inventorying the resource for which it was responsible, Leopold found himself exploring the forest's wildlife, establishing working relationships with the area's sportsmen, and joining with them to establish a more effective state game administration.[34]

Whether the new national forest administration liked it or not, it was thrust into not just managing wood production but also the western world of game and hunting, much as it had been forced to address the livestock grazing issues it had inherited. But its response to wildlife on the one hand and sheep and cattle on the other differed markedly. While it chose to play a significant role in the latter, it chose to distance itself from the former. Leopold responded to the agency's approach to game and hunting not as an unwelcome irritant but as an opportunity. In doing so, he took his responsibilities far beyond his assigned wood production duties.[35]

In these efforts, Leopold's immediate superiors gave him encouragement and looked upon his work as constructive within the broad outlines of agency objectives. But this positive response did not go beyond his district to the leaders of other districts or to the agency's headquarters in Washington. Assistant Forester Leon Kneipp replied to Leopold's wildlife inquiries with the negative remark that "existing statutes do not recognize wild game as a forest product."[36] The fate of Leopold's wildlife interests illustrated the way in which imaginative innovators in the agency were restrained by the management choices shaped under the direction of Chief Forester Gifford Pinchot. Pinchot had been one of the early leaders of the Boone and Crockett Club, but in his mind, wildlife reserves were not to be one of the new agency's responsibilities. As a result, the Forest Service evolved in its early years without accumulating experience in wildlife management and with a studied rejection of its importance.

Within the new national forest agency, grazing and wildlife somewhat competed for administrative attention, and the primary commitments to grazing helped to reduce wildlife to a more secondary role.[37] Agency staff spent a large amount of time on grazing matters, seeking to establish a more organized system of allocating use of the range and to reduce grazing overall in order to protect and restore range forage. In so doing, they sought to establish closer working relationships with the stockmen and, toward that end, to see range policies from that public's point of view as well as their own.

Cooperating with the stockmen, Forest Service personnel helped to eliminate predators that killed sheep and cattle and thus were a threat to a viable livestock industry. During his time in the Southwest, Aldo Leopold himself urged ongoing efforts to protect wildlife as well as livestock by vigorous antipredator campaigns, a view that he continued to hold for many years. More difficult to work out, however, was the competition between livestock and game for scarce forage, which inevitably led sheep owners and cattle raisers to have some hostility to the notion of game preservation. Such pressures, along with agency officials' tendency to sympathize with the stockmen, meant that game was to take a backseat in the ongoing readjustment of range-use allocations.

Such predispositions on the part of agency officials resulted in the reassignment of the agency's few game staff to the Division of Grazing. There they remained until 1937, when the Forest Service formed a separate Division of Wildlife with its own administration.[38] Only then could wildlife advocates within the Forest Service see the serious beginnings of an independent wildlife program for the national forests.

Wood Production

Despite the preoccupation of the Forest Service in its early years with grazing management, it forged ahead with what the agency considered to be its primary responsibility: wood production. In this it started from scratch, so to speak. It deployed many a young man with forestry training, along with men experienced in woodland affairs, into the western national forests simply to find out what was there. They surveyed the public lands, identifying geographical features, forest and range lands, rivers and streams, and conditions relevant to the growth of trees. Their primary interest was in the "growing stock," that is, the species of trees, their age and condition, and the amount of wood fiber they contained.

Details about their activities are most readily available in the biographies of individuals involved, such as Curt Meine's biography of Aldo Leopold, whose initial employment was on the Apache National Forest in northwestern New Mexico, or in the history of particular forested areas, such as Nancy Langston's account of early Forest Service work in the Blue Mountains of northeast Oregon. The task was simply to find out what was there, and the most important work was to "cruise" the land and generate a detailed inventory of its timber resources.

All of this was inspired by a mission to establish "sustained-yield" wood production. The message of this mission was that "timber is a crop"—that trees for wood could be subject to the same sequence of planting, tending, and harvesting as was done for the usual battery of agricultural crops. Moreover, trees could be managed in such a way that the result could be a continuous crop if what was removed periodically was no more than what was grown during the same period. This was a far cry from past practice, in which trees were harvested without providing for new growth.

The firmest expression of this new practice, which some considered ideal, was that in harvesting wood one should be alert to what was left as much as to what was taken. This was a call for a long-range view to wood production rather than one limited to momentary benefits or the notion of "timber mining," a view that likened timber extraction to mineral extraction. Thus it was that the U.S. Forest Service was to be a shining example of long-range perspective in forest management, a perspective more attuned to the long-term needs of society for wood products amid the more prevalent industry preoccupation with short-term economic gain in forest removal.

This new vision in wood production and forest management was called silviculture. Its objectives, which the personnel of the new agency carried into the field, were to liquidate the existing forest and replace it with a more intensively managed one that could be described as a "regulated" forest.[39] "Regulation" in silvicultural language meant a forest that was fully shaped from planting to harvest through careful and continuous management of the trees.

The existing older forests of the reserves were thought to be relatively useless, experiencing very slow growth and capable only of being harvested rather than serving as a model for the future. This approach to forest management—liquidating the old in order to begin the new—immersed the emerging forestry profession in the United States in the silvicultural concept and constituted a primary contrast to European forestry; in Europe there were no old forests to contend with. In America, "scientific" forest management could not begin until the old forests were out of the way.

In his history of the profession, Clepper emphasized the difficulty of developing a "regulated" forest while simultaneously getting rid of the older, less productive one. "One must understand," he recounted, "that the commercial forests consisted almost entirely of overmature, old-growth stands. From a silvicultural standpoint, the requirement was to cut heavily in order to remove the timber that had reached or passed economic maturity or that had slowed in growth to the point of little or no net increment. Because of the need for heavy cutting, an initial depletion of the forest capital resulted. But this in turn was followed by marked acceleration of the second growth, permitting silvicultural practices."[40]

Thus, silviculture in the national forests began with a vigorously negative attitude toward "old growth." Management strategies looked to its rapid liquidation so that the new "regulated" and more highly productive forest could replace it. The first step in a program to counteract a future "timber famine" was to increase rather than retard cutting of the existing forest. This rejection of the value of old-growth forests and invitation to wood production regulation in increasingly elaborate detail shaped a fundamental vision about the application of scientific forestry for the years to come.

These guiding objectives in forest management, however, soon faced economic circumstances in the wood production industry that brought about some significant modifications.[41] The first was the fact that the industry was selective in what it was willing to purchase. It wanted the larger trees, which were the most profitable, but it balked at the agency's desire to include in the timber contracts species that it wished to eliminate but for which there seemed to be little market and its requirement that the purchaser clean the harvest area of slash, brush, and tops in order to facilitate the planting of trees for the new forest. Thus, the young agency faced the task of negotiating contracts in which its objectives and the objectives of the market came into conflict.

This challenge was only the beginning of a process in which the market exercised limitations on the agency's silvicultural ambitions in varied ways over the years to come. The agency was often forced to accept terms established by the wood buyers in order to entice them to carry out agency objectives.

Equally significant in modifying the agency's silvicultural objectives was the evolution of its tendency to think of the wood harvest and sale program in terms of how it would support mills that would purchase the harvested trees.[42] The market held implications not only for timber buyers but also for the mills. The mills represented local employment and local economies, which often prompted the agency's sales specialists to harvest faster than the sustained-yield schedule called for. This premature cutting then led to predictions that wood supplies would run out long before growth would permit a new round of cutting.[43]

Much of the spirit of a new approach to wood production was a reaction against the impact of rapid timber extraction on timber-dependent communities. The long-standing "cut and run" practices of the private timber industry had pulled the rug out from under the community's economic livelihood. An industry based on continuous sustained-yield production would provide greater stability. Thus, even the appearance that modifications in the agency's sales policies might shake the community's economic base tended to reinforce the idea that community stability did not always square with the requirements of the silvicultural circumstances that the agency was attempting to apply. It was relatively easy for "sustained yield" to be reinterpreted, as it was later in the twentieth century, to mean "sustained communities," which might well lead to cutting levels not consistent with sustained-yield timber production. These long-term implications could be seen even before World War I, in the earliest years of the agency's sales policies.[44]

Physical constraints of topography and soil productivity also led to modifications in the agency's silvicultural ambitions. Throughout the history of the logging industry the problem of access and transportation from forest to mill had played a crucial role in exploiting timber resources. At first harvest was limited to areas near waterways in which logs could be floated to the mills. Then the construction of narrow-gauge railroads made it possible to cut timber in lower elevation stands if the terrain was not too steep.

In later years, logging roads for trucks provided access to higher elevations, but timber on steeper slopes required helicopter operations.

Throughout its history, the Forest Service had to lower the price it charged for timber so as to make its contracts attractive to buyers. In the earliest days of the national forests, this practice set a profit margin for buyers that would cover the cost of constructing a logging railroad to transport logs to a mill.[45] Using the sale of trees to finance their transportation greatly complicated the financial circumstances of the agency, particularly its claim from the earliest days that it could make sustained-yield wood production "pay." To adjust the operation's finances to avoid identifying a cost as a cost was always tempting.

Soil productivity issues also constrained the agency's ambitions. National forest lands varied considerably in their natural productivity because of geography, with prolific growth in the Pacific Northwest, where rainfall was abundant, and meager growth in the intermountain region between the Sierra Nevada and Cascade ranges to the west and the Rocky Mountains to the east. Private logging outfits had already taken up the most productive lands, just as they had already acquired those that were most accessible with the least cost.

The less productive or more marginal lands greatly thwarted the agency's silvicultural ambition of clearing away the old forest to establish a new one. Over time, many argued that the agency should log only lands where adequate regrowth could be anticipated, which would seem to be a logical extension of the "sustained-yield" objectives. The agency responded that the test of desirable logging should not be future productivity but current need for the lumber, a view that seemed to be a further adjustment of silvicultural goals to the realities of the marketplace.[46]

Aesthetics

Natural beauty was one of the characteristics of forests that regularly attracted the public's interest in their protection. This interest was expressed on many fronts and spread to larger numbers of the public as their recreational and leisure-time experience with forests grew, from the initial cases of the more wealthy who established their own individual forest "camps" in the Adirondacks to the public at large after automobile travel increased their access to forests after World War I.

Persons who had an interest in forests because of their natural beauty, however, came into conflict with those who looked upon them as a material asset. For the emerging group of professional foresters, the aesthetic qualities of forest lands played no role in their scheme of management. They reacted sharply against such views and looked upon them as detrimental to the success of effective silviculture.

These utilitarian views were expressed with a similar degree of vigor by the two most widely known spokesmen for scientific forest management: Bernhard Fernow and Gifford Pinchot, particularly with regard to the management of the Adirondacks forest lands.[47] For Fernow, the "aesthetic factor" had considerable importance because it involved the central element in the Adirondack residents' objections to his management of a thirty-thousand-acre tract of forest land that Cornell University had acquired as a demonstration area in connection with the university's forestry school, which Fernow headed. Fernow dismissed the aesthetic interests of New Yorkers as merely an attempt to create "a luxury forest for the rich," that is, people who did not recognize the value of replacing "the old decrepit natural forest with a new, more valuable forest." Recognizing the need to follow the public preference for forest aesthetics in managing the college tract, however, he called for two-hundred-foot-wide strips of forest to remain around cut areas so as to screen all highways and other travel corridors from logging operations.[48]

To Pinchot, public interest in the natural beauty of the nation's forests was potentially a major roadblock to sustained-yield management of the forest reserves. He was thus persistently unwilling to bring aesthetic values into forest management, and it was left to the agency in the 1920s to institute a program to protect the natural aesthetic quality of travel corridors. On the whole, however, the views of both Fernow and Pinchot were not just their own; they reflected rather fundamental views of those who expressed a primary interest in "scientific forest management." The attitude of Pinchot and the early Forest Service toward aesthetic objectives in forest management was as vigorously negative as their attitude toward wood production was positive. Whereas grazing was a circumstance the agency could not avoid and watershed and wildlife objectives were minimized by being dealt with quietly and without much debate, aesthetic objectives were rejected in often overt ways.

Scholars have dealt with such matters in the early history of the national forests largely through discussions of ideological conflict, the "preservation" versus "use" debate with the conventional "parks versus forests" set of opposites. As a result, historians have effectively obscured the role of aesthetic objectives, despite the fact that the issue cropped up in a number of national forest management circumstances. This has even prompted scholars to give prominence to the issue of constructing a dam in the Hetch Hetchy Valley in the Yosemite National Park as a part of the history of the national forests when it was actually a part of the history of the national parks. In this phase of forest history, the incremental actions in the expression of values and the agency's institutional development provide far more insight into the role of aesthetic objectives in the history of the Forest Service than does the more prevailing focus on alternative ideologies.[49]

Although the statutory authorization for the agency did not specify aesthetic objectives and they were not explicitly identified even in the Multiple-Use Act of 1960, the Forest Service did, in fact, implement them when it chose to do so and, in turn, rejected them also at its discretion. In the years after World War I, the agency began to address aesthetic objectives as circumstances gave rise to their apparent importance in achieving the agency's overall goals.

The aesthetic issue most widely debated in the early years of the Forest Service involved the administration of the emerging group of national parks. Advocates of harvesting timber from the national forests seemed to recognize well the distinction between logged forests and unlogged parks simply by the vehemence with which they emphasized that the national forests were not parks and thus were not to be thought of as aesthetic areas. Gifford Pinchot was especially insistent on this principle and declared so repeatedly. Yet in the minds of many advocates of the forest reserves, the distinction was not so sharp.[50]

Some of the most vigorous supporters of the reserves, such as the American Civic Association and the General Federation of Women's Clubs, combined both commodity and aesthetic objectives in their thinking. While their objectives differed, the various supporters of the reserves believed that forested areas could be administered for both wood production and aesthetic objectives under a common administrative arrangement. Moreover, proponents of specific forest reserves often thought of them as potential parks or wildlife refuges and recommended reserves with this in mind. It is significant that several states have combined administration of their state parks and state forests with no apparent compromise of the quite different objectives of each.[51]

The inclusion of parks in the early national forest policy debates was reflected in proposals of the National Academy of Sciences (NAS) in 1897 regarding the significance of the reserves. The NAS recommended that two new parks should be carved out of the Rainier Forest Reserve in Washington and the Grand Canyon Forest Reserve in Arizona.[52] Pinchot, however, was vigorously aggressive in his campaign to exclude parks, with their "no timber harvest" management policy, from forest administration. He proposed amendments to a number of bills providing for specific parks so that, in revised form, they allowed timber harvest in the parks. In 1905, he advocated that the parks be transferred to the Bureau of Forestry so that timber could be cut in them, a proposal that was effectively opposed by Rep. John Lacey (R-Iowa), chair of the House Public Lands Committee, who favored transfer of the forest reserves but not the parks.[53]

Pinchot also opposed allocating some of the sequoia groves in the Sequoia Forest Reserve for an expansion of Sequoia National Park, a proposal that had come from the California branch of the General Federation of Women's Clubs. He declined to support the proposal, offering the disingenuous claim that it was not appropriate for the chief of the Forest Service to take a position on legislation pending in Congress.[54] Pinchot sharpened the issue between Forest Service and park advocates during the Governors' Conservation Conference in 1908. He gave only token support to parks, aesthetic issues, and even wildlife in the conference agenda, which he had shaped, and in the agendas of the conservation conferences that followed.[55] He even declined to allow park advocates such as John Muir and the head of the General Federation of Women's Clubs to speak at the 1908 conference.[56]

One of the more instructive episodes in this series of debates over aesthetics and parks was Pinchot's relationship with Enos Mills, which is described in Alexander Drummond's biography of Mills. A Coloradoan and early advocate of both the reserves and the Rocky Mountain National Park, Mills had close ties to the General Federation of Women's Clubs and especially its Colorado branch during the effort to establish the park. He became a well-known speaker, and after he was recommended to Pinchot as an able spokesman for the national forests—in his talks, Mills indicated no preference for forests as sources of wood versus natural beauty—Pinchot hired him in January 1907 to lecture throughout the country in support of the national forests. Yet Mills was given no role at all in the Governors' Conservation Conference of 1908, and no leader from the women's federation was invited to speak. Early in 1909, Mills was dropped from the agency's employ. The episode was well known to the advocates of aesthetic values, and they interpreted it as part of Pinchot's firm rejection of aesthetics as an integral part of national forest management.[57]

From the conference in 1908 forward, park advocates no longer held out the hope that the Forest Service would provide an acceptable administrative home for parks, so they devoted their time and energies to establishing a separate National Park Service. Pinchot opposed this move as well but did not succeed in squelching it.[58] The view of the Forest Service that aesthetic objectives were not a part of its mission only prompted park advocates to concentrate their energies on the work of the National Park Service, which led to a sequence of proposals by park advocates to transfer lands from the national forests to establish new national parks.

The Agency and Congress

Congress was the ultimate governing authority under which the Forest Service administered the national forests, and the Transfer Act of 1905 was the legislation under which the agency began its operations. Congress's supervision of the Forest Service took the form of an annual scrutiny of the agency's finances as part of the budget process. It is rather remarkable that historians have given little attention to this financial supervision. Not until more than a half century after the agency's beginning were these financial arrangements examined carefully as a window into the larger question of the relationship between the agency and Congress. It was Robert Wolf, a trained forester working for the Congressional Budget Office, who undertook the study in order to understand the history of the agency's finances. Those financial operations became the subject of considerable debate in the late 1970s.[59]

When Gifford Pinchot left France after studying forestry at l'École Nationale Forestière, his mentor urged him to return to the United States and take up work to "make the forests pay."[60] But in his first forestry job, managing the 3,891 acres of low-grade hardwood second-growth timber land on the Biltmore estate in North Carolina, Pinchot encountered some economic roadblocks that were more than sobering. It seemed almost impossible to assume the cost of making improvements in the timber stand and break even financially.[61] Clepper, long associated with the Society of American Foresters, writes that "many well-qualified foresters would be defeated by this problem during the following seventy-five years, for the situation did not change until pulpwood markets became available."[62] From his experience at Biltmore, Pinchot himself observed that the economic output of the Biltmore forest was unfavorable.

Despite this experience, when Pinchot was seeking to persuade Congress to transfer the forest reserves from the Department of the Interior to his agency, the Bureau of Forestry, he sought to give weight to the proposal by arguing that he could "make the forests pay." Under the jurisdiction of the Interior Department from 1897 to 1904, reserve management had generated $203,100 in revenue but had $1,605,700 in costs. Pinchot argued that he could do much better and that he would not only generate income from the reserves but also enable them to serve as examples to private forestry for the value of scientific forest management.

Pinchot sought considerable agency freedom from the usual legislative oversight of agency finances. He asked for, and received, authority to retain agency proceeds and spend them on the "protection, administration, improvement, and extension of the Federal forest reserves," a request that Congress granted but limited to five years. This action placed the Forest Service under a considerable challenge to "make forest management pay," which led to continued congressional scrutiny of the agency's finances.[63]

In his testimony at congressional committee hearings after the transfer of the reserves, Pinchot repeatedly claimed that the forests were either currently profitable or soon would be. In his financial reports, according to Wolf, "he used selective cost data and juggled his authority to spend timber sale receipts. A hard look at the Forest Service under Pinchot reveals that the agency and the timber program continued to lose money."[64]

After running deficits for several years, Pinchot explained that they were caused not by normal management practices but by inadequate management infrastructure. To obtain capital for agency development, he requested $2 million as a loan (he had originally planned to request $5 million, but, sensing congressional resistance, he reduced it to $2 million), which the agency would repay, with interest, at $500,000 per year beginning in 1917. Congress rejected this proposal, granted only a $500,000 direct appropriation, and canceled the provision in the 1905 law that permitted the agency to use revenues to cover agency costs.[65]

Amid this ongoing tussle between the agency and Congress, Pinchot himself began to turn away from the argument of "making the forests pay." In his 1909 testimony to Congress, he requested that the agency not have to be preoccupied with getting "every cent we can out of the national forests in order to make them self supporting," and he began to develop arguments to justify Forest Service affairs in ways other than financial profitability. One argument was that agency management was satisfactory because the national forests were rapidly increasing in value. Another was that an increase in income was a satisfactory measure of financial success; relying on a single measure suited his purposes because it obscured costs and financial losses. He also began to place more emphasis on the nation's coming "timber famine" and the need "to do the best thing for the forests and the best thing for the country at large, instead of fixing my main attention on revenue."

The continued weakness of his financial claims for national forest management led to Pinchot's increasing focus on private forest management in the years after he left the agency in 1910.[66]

The tussle between Congress and the agency over financial matters continued in the years from 1910 to 1920. Legislators, for example, asked the new chief, Henry Graves, why annual costs, which were $350,000 when the forests were managed by the Department of the Interior in 1904, had risen to $5.5 million in 1910. Graves sought to improve the agency's financial position primarily by increasing timber sales even in years in which there seemed to be an overabundance of wood on the market. He also sought to divert attention from the agency's overall financial condition to accounting on the basis of individual forests or what were called "local operating costs." While some forests did not "pay" then and eighteen would never pay, he argued, by 1923 more than 130 forests were able to cover those local costs. And he predicted that by 1928, receipts for the agency would cover all costs, including those agency activities that were not intended to produce a profit.[67]

According to Wolf, "In sum, the first fifteen years of Forest Service timber sales were years of optimism, an optimism powerful enough to obscure mounting evidence of widespread unprofitability."[68] In 1920, the agency's report of its financial activity from 1909 to 1919 showed that expenditures exceeded receipts by $28.5 million. The agency reports may have justified its finances in terms of an increase in revenue alone or a promise of profitability in the future, but on the basis of those results for the decade, the agency could not say that it had used its appropriations to make the forests pay.

Who Makes the Decisions?

In their early years, the forest reserves and the national forests were heavily dependent on decisions made by the president and the executive branch of government.[69] The reserves had been established by executive order and, aside from the act of Congress that transferred their management from the Department of the Interior to the Department of Agriculture, there was little legislation through which one could detect a significant amount of public support for reserve management. One observes a kind of cat-and-mouse game between Congress on the one hand and President Roosevelt and Gifford Pinchot on the other. It was a game that culminated in the act of Congress in 1908 to prohibit further executive declarations of reserves in the northern states and was followed almost immediately by even more executive orders of that type and in those same states between the time that the legislation passed and Roosevelt signed it.[70]

It is difficult for the historian to determine the degree of popular support or lack thereof for the agency throughout the country as a whole.[71] This unknown quantity stands in contrast with quite visible support for the protection of game in the Yellowstone National Park Act of 1894, which established a firm system of enforcing regulations prohibiting poaching in the park. The widely publicized killing of bison in Yellowstone for their hides and trophy heads led to an outpouring of public sentiment that enabled that act to pass with little opposition, a far cry from the constant skepticism expressed in Congress on issues pertaining to the national forests.[72] The Lacey Act of 1900, which prohibited the interstate shipment of wildlife taken illegally, and the Antiquities Act of 1906, which provided for protection of archeological and cultural sites of historical interest, also passed with little opposition but less public outcry.[73]

All three of these laws, it should be noted, originated in the Public Lands Committee and went through the normal legislative process under the guidance of committee chair John Lacey of Iowa, in contrast with the major pieces of forest reserve legislation, which Congress passed when inserted at the last minute in other, unrelated legislation. John Lacey's quiet and effective leadership was in sharp contrast with the more flamboyant public activities of Gifford Pinchot.[74] Amid these observations, one has difficulty in identifying significant active public support for the forest reserve and national forest program. Despite the willingness of historians to make statements about that support, few have made a convincing argument that it was widespread. The most visible organized backing came from the American Forestry Association, formed in 1876, which both advocated extensively for the reserves and gave strong support to the management structure that Pinchot created in the early years of the twentieth century. However, membership in the AFA was limited and consisted primarily of professional leaders.[75]

Far more extensive public support for the national forests in the late nineteenth and early twentieth centuries came from sportsmen and those with aesthetic interests in the forests. This was a clientele that Pinchot alienated, in contrast to the group of state and national leaders whose support he sought to further the development program outlined in the Governors' Conservation Congress of 1908.[76] The Forest Service was part of a complex of programs that included the Reclamation Service, which was involved in bringing water to the desert West; the reservation from sale and permanent federal management of underground oil and gas holdings and water power sites; and the promotion of navigation on the nation's inland waterways, all of which were described as elements of the Roosevelt administration's "conservation" program. These were thought to be resources of considerable development potential that should be brought into use by professionals with expertise in both efficiencies of operation and technical knowledge.

The Forest Service shared this enthusiasm for efficiency and expertise in resource management because its main objective was not just "sustained yield" but also management that would benefit the nation by reducing waste to a minimum. The new agency's decisions were to be made not in response to public desires but according to the views of experts.[77]

The U.S. Forest Service, therefore, developed under the protection of the president and the guidance of an elite corps of technical leaders. It had little in the way of a visible and active public clientele. In fact, the choices the agency made in its management direction, largely under the influence of Gifford Pinchot, had alienated the very element that had provided a favorable public climate in the early stages of the reserves and for the transfer from Interior to Agriculture. The most extensive segment of this public, those with wildlife and aesthetic objectives, who were either neglected or excluded in the new management scheme of things, simply went their own way and became associated with other resource programs such as the development of a system of national parks and wildlife refuges. Watershed users with a direct interest in watershed protection also found little in the new agency to attract their support.[78] This divorce from the group of potential forest supporters was fostered by the limited attention Gifford Pinchot allotted to their issues at the Governors' Conservation Conference in 1908 and its successor organization, the National Conservation Association, where wildlife and aesthetic objectives were noticeably absent and watershed issues were actively subordinated to water development objectives.

At times Pinchot seemed to recognize that there was limited public support for systematic forest management.[79] The problem of fire detection and suppression aroused some public interest, and the agency's desire to form a disinterested civil service rather than provide a haven for patronage appointments attracted admiration. When Pinchot sought to arouse congressional support for the transfer of the reserves, he seemed to understand the need to "build a fire" under the legislators. For that purpose he organized a "forest congress" composed primarily of the nation's prominent leaders, but this did not involve any development of a continuing constituency with which the agency could work to solidify public support in the face of westerners' continual attempts to undermine federal control and management of the reserves.

After the Governors' Conference, which actually did cover resource objectives beyond simply forest management (in fact, at the conference, forest matters were greatly overshadowed by the attendees' preoccupation with water development), Pinchot believed that there perhaps was need for an active public clientele, so he organized the National Conservation Association for that purpose. This effort still did not include overtures to the more extensive potential constituency of wildlife, watershed, and aesthetic objectives advocates. The new organization was, in Clepper's words, primarily a "one-man show," which Pinchot fostered largely because he had failed to influence the American Forestry Association to his satisfaction. Pinchot fully expected that the new organization would rapidly attract some fifty thousand supporters. Yet within a few years subscriptions to the organization's magazine, *American Conservation,* totaled fewer than two thousand and the publication was dropped.[80]

The Forest Service entered the post–World War I era with active western opposition and an almost nonexistent national clientele.

The Silvicultural Legacy

Despite its somewhat tenuous political existence, in the first two decades of the twentieth century the U.S. Forest Service shaped for its times a rather formidable management system. The agency gained enormous respect for its nonpartisan approach to wood production and grazing affairs; its emphasis on technical training for its employees; its requirement of civil service exams for recruitment; its emphasis on building a program based on knowledge and science about trees, their cultivation, and their products; and, what seemed surprising to many, its ability to translate the standards set by the agency in Washington, DC, into action by its employees on the ground. This, as historians have rightly argued, established the beginnings of an agency with high professional standards for impartial resource management. This silvicultural imperative provided a permanent legacy that carried the Forest Service through the tumultuous years of the twentieth century.

The same commitments to the development of management efficiency, however, involved a similar intensity of commitment to limited management goals. These involved choices that narrowed a wide range of public objectives implicit in the early years of the reserves, from the initial legislation of 1891 onward, to a more limited set of objectives by the time of the transfer in 1905. Pinchot was consistently firm in making clear the choices that he had made. For example, when writing to Alaska's territorial governor, John G. Brady, about management for the Alaskan reserves then being established, he stated simply, "Forest reserves are not parks or game preserves."[81] While in the latter part of the nineteenth century the public had thought of the reserves in terms of wildlife, watershed, and aesthetic objectives as well as wood production and grazing objectives, by the time the forests were under the control and direction of Gifford Pinchot the first three objectives had either been excluded or subordinated in agency strategies in favor of the last two. The agency had not only greatly narrowed its goals but also its clientele and had bequeathed to its future managers an institution not only of enormous skill and competence but also of narrowed vision.[82]

This limited view can be contrasted with the broader range of forest objectives that were being carried out in New York. Andrew Rodgers, in his biography of Bernhard Fernow, reports that a succession of New York governors —David Hill, Roswell Flower, Frank Black, and Theodore Roosevelt—"in strong language ... [each] had said that the state-owned forests might and should provide a large source of revenue to the state, as well as protect water supply and provide summer homes and sanitariums and pleasure resorts."[83] The state's leaders were expressing a far more comprehensive set of forest management objectives than Pinchot had in mind for the Forest Service.

Since the Adirondacks did not provide Pinchot with the opportunity he sought to practice "scientific forestry" and the Biltmore experiment had been equally unproductive, he turned to the Chippewa Forest Reserve in Minnesota for another possible opportunity to put his theories in practice. This forest had an unusual history, described in some detail in Pinchot's autobiography, *Breaking New Ground*.[84] The Chippewa Indian Reservation was proposed initially by the Minnesota state forest warden as a national park. The proposal was taken up eagerly by the state's Federation of Women's Clubs, a delegation from which went to Washington, DC, in 1902 to present their proposal to Rep. Robert P. W. Morris, who represented the area in Congress. In the course of negotiations, the park proposal became a proposal for "scientific forest management," a transformation in which it is reasonable to suspect that Pinchot had some influence since the legislation provided that the forest reserve would be managed by the Bureau of Forestry, which Pinchot administered.

After Congress approved the reserve proposal in 1902, Pinchot enthusiastically set about applying his version of "scientific forestry." His discussion of some of the practical problems entailed in negotiations with private forest firms illustrates that this science was far more than a theory of "sustained yield." In his contracts with wood purchasers, he insisted that they dispose of the wood debris from harvesting, a requirement that they resisted with great determination on the grounds that they could not absorb the cost. Management of the Chippewa reserve provides details of the model of wood production forestry for which Pinchot was searching and which he hoped to apply to the federal forest reserves if and when they were to be transferred to the Department of Agriculture.[85]

The Pinchot legacy set the stage for much of the future of the Forest Service, especially the ongoing tussle with those who had earlier been excluded from agency objectives and continued to demand inclusion in the Forest Service's program, which was a major theme of the years between 1920 and 1960. Their "admission" was successfully formalized in the Multiple-Use and Sustained-Yield Act of 1960, and in subsequent years their active involvement in agency affairs gave rise to tension over the role these wider objectives were to play in the administration of the national forests.

The Forest Service evolved not only under the protection of the executive office of the president; it also was heavily shaped by the culture of expertise that surrounded the natural resource programs of the Theodore Roosevelt administration. That professional culture did not spring from the influence of a popular movement but was more related to the thinking of technical experts in water and land issues. The resistance of the agency's corps of committed wood production and grazing civil servants to the larger role of those publics excluded from Forest Service affairs colored the agency's early history. Its image was not just that of a perfected managerial system but of a tenuous relationship to the public's changing attitudes as to the importance of forests.

3 EVOLUTION OF AN AGENCY CLIENTELE, 1920-1975

DURING the years between 1920 and 1975, the public circumstances surrounding the Forest Service evolved markedly. These new circumstances did not arise from legislative innovations or innovations from within the agency or the profession but from the broader changes in society. The increased use of automobiles made the forests more accessible to hunters, campers, and hikers, whose activities impinged on forest management. The wilderness movement, reflecting, in part, the increase in outdoor recreation but especially the steadily increasing numbers of people who expressed interest in the aesthetics of the nation's wild areas, grew steadily. At the same time, the timber and livestock industries imposed increasing demands on the agency. These demands were primarily fueled by pressures to increase and even maximize production, and after World War II, these demands did not subside but instead increased even further.

Such changes meant that greater demands arose from many and varied quarters, and the agency had to meet these expectations for uses other than wood production, such as outdoor recreation, hunting, and aesthetics. There were two major results: the first consisted of demands that these various uses be given statutory authorization and the second, that agency management indicated that it was capable of maximizing "outputs" to satisfy all. The first received recognition in the Multiple-Use Act of 1960, and the second was addressed in the authority given to the agency in that act to determine the balance and level of usage that would satisfy the users who now took a growing interest in the agency's management.

Historians have often identified World War II as a turning point in national forest history on the grounds that it was the wartime demand for lumber that accelerated a major change in the role of the Forest Service—from custodian to manager.

This view seems plausible if one gives primary emphasis to wood production. However, the war retarded rather than accelerated recreation, wildlife, and wilderness activities—the steadily emerging "other" uses. The war interrupted the growth of recreation activities in the forests that had occurred in the 1920s and 1930s; that growth resumed its pace after the war. With respect to wildlife objectives, the war halted growth in those agency activities, which remained slack for many years. And the war prompted the agency to take steps to retard previous wilderness gains by reevaluating the boundaries of protected areas, a strategy that involved releasing some acreage for wood production. The role of the war in agency affairs was thus mixed, advancing extractive resource objectives but interrupting for a few years those devoted to recreation, wildlife, and wilderness.

Agency-Industry Rapprochement in Wood Production

Working relationships between the agency and the wood products industry were implicit in the respective objectives of agency and industry. The agency objective was ensuring the nation's long-term wood supply, and the industry's concern was the immediate profit to be earned from wood harvest, processing, and sale. But the presumed working relationship had already experienced a rocky history, one that went through a number of stages over the course of the twentieth century.

In the early days, the industry looked upon the two parties in the relationship as carrying out distinctive but integrated objectives: the industry would cut and market standing timber, and the Forest Service would acquire cut-over lands and grow a new crop. As late as 1920, for example, the National Lumber Manufacturers Association, formed in 1902, was still urging government purchase of "much larger areas of permanent forest land than they now possess." It also announced that "if private owners refuse either to sell for such purpose, or to take reasonable steps themselves to keep in timber crops, any deforested land competently classified as suitable chiefly for forest growing and not suitable for agriculture, Government and states should be permitted to condemn and pay for it at prices comparable to those paid in voluntary transactions."[1]

The Forest Service, however, with its broader concern for long-term supplies, believed that industry's obligations went further. In the agency's view, the forest industry should be responsible for the future as well as the present, should replant cut-over land, and tend trees to grow for another round of harvest. Thus, as the wood-production specialists in the agency shaped not only their program for the country but also the management of its own lands, it began to be skeptical that the nation's objectives could be met by the industry's more limited perspective.

Gifford Pinchot was the most outspoken proponent of this skepticism. As early as 1909, he set forth in a major conservation publication his belief that private industry could not take up sustained-yield wood production. Because standing timber was still relatively cheap and its price low, the industry could not afford to finance the costs of wood production, and its activities could not compete with other investment opportunities. Also, the potential for timber destruction from fires often persuaded forest owners to rush their timber to market. The desire for quick returns precluded the long-term investment required for sustained-yield production. These industry concerns lingered for a number of years, but in the 1920s, they were translated into firm proposals for public regulation of private forests.

Those proposals ignited a controversy between the agency and the industry that lasted for more than three decades and continually disrupted potential collaboration between the two sides.² Every chief forester from Henry Graves (serving 1910–1920) to Richard McArdle (serving 1952–1962), save perhaps William Greeley, who succeeded Graves in 1920, advocated some sort of regulation. Some would have limited it to fire detection and suppression rather than cutting practices; others, such as Ferdinand Silcox, chief between 1933 and 1939, proposed that to prevent a potential "timber famine" disaster the agency should not only grow the trees but also process and sell the timber.

The issue waxed and waned. But suddenly, after World War II, it disappeared from the agency agenda. The less than dramatic end to the controversy came in 1952. In his last report as acting chief, Lyle Watts in 1951 was still advocating regulation, but in the first report of his successor, Richard McArdle, in 1952, the issue was simply ignored and it was never again mentioned.³ Soon a rapprochement between the industry and the agency began to emerge, reflecting a joint commitment to advancing wood production as the agency and industry's primary mission.

Several changes within both the industry and the agency facilitated this cooperation. Within the industry, several firms had begun to think of their forests in more long-range terms and had taken up planting for a new crop rather than simply harvesting the old.⁴ The number of such firms was small even into the 1950s but sufficient to create the image that, in this limited way, some were willing and able to take on these longer-term responsibilities. This new industry interest in the long term was facilitated by a major shift in the market from lumber to pulp and paper, which greatly shortened the time span of the "long term."⁵ The National Lumber Manufacturers Association had to yield its role as the main voice of the industry to the American Forest and Paper Association. Pulp for paper production could be grown with much shorter rotations than for lumber, and the surge in pulp and paper gave rise to a forest industry interested in growing trees more rapidly, with more intensive cultivation. Stockholders in these new ventures would be able to realize income on their investments more rapidly than was the case with a lumber-manufacturing industry.

The changing conditions of the market also led to changes in the perspective of the Forest Service.⁶ The influence of the market on the agency had been subtle but profound, causing it to modify much of the silvicultural program that had inspired its earlier activities. The agency continued to have difficulties with purchasers who resisted the contractual requirement that they conduct forest management operations, which generated costs rather than income. At first these management operations involved wood products such as slash and species the agency wished to get rid of but had no special funds earmarked for that purpose; it simply included responsibility for that management objective in the purchaser contracts.

Another problem, the cost of transporting timber to market, grew as the use of trucks replaced railroads and necessitated construction of forest roads to access desired harvest areas. The agency met this expense simply by subtracting the cost from the contract bid so that roads were, in effect, built by proceeds from the sale of timber itself.[7] To justify such practices, the agency would allocate the cost savings to uses other than timber, such as recreation and wildlife, which produced no specific agency income.

Much of the agency's wood production program was shaped not by the general market but by the specific markets represented by the mills that its timber would supply and by the accompanying jobs, which were often the economic life blood of communities that depended on the mill.[8] Because of these vital relationships, the Forest Service was inclined to bend the harvest limitations that a conservative silvicultural program called for so as to protect a community's mill and jobs.

The obvious implications of such adjustments for the agency's long-term mission led to several larger policy adjustments. First was the guarantee of a "nondeclining even flow" of timber for the mills and the community and second, the idea that the meaning of "sustained yield" should be modified to mean "sustained communities," that one of the agency's most important objectives was not to provide for a sustained production of timber but to sustain the economic life of communities by providing for a stable economy.[9] Amid such pressures, the requirements of "sustained yield" often could be compromised by community economic concerns.

Rapprochement between the industry and the agency had been facilitated even during the early years of the dispute over public regulation of private forests by the agency's approach to two issues, fire and replanting, which emphasized state action that would be financed, in part, through federal grants.[10] Industry was more interested in the fire question, since fire was a major threat to its forest holdings and increased the speculative nature of its investments. Private timberland holders had been a major influence in urging the states to take up fire detection and suppression activities. But the states were lethargic on this score.

The Weeks Act of 1911, which provided for acquisition of national forest lands in the East, allocated funds that the Forest Service could grant to states to support fire control programs if the states made matching grants. Some states began to take up this offer, and in the Clarke-McNary Act of 1924 the program was expanded.[11] In a program structured in a similar fashion, federal funds were made available for replanting cut-over forest lands, which were used primarily to set up state-run tree nurseries that provided seedlings at a low cost. These innovations came during the thick of the controversy over regulation and have been understood by historians as an attempt to reduce the intensity of the agency-industry debate.

These efforts at fire control and replanting were given even more extensive support by the Civilian Conservation Corps program during the New Deal.[12] One could readily observe the close connection between the former state programs and the new CCC program through the political career of Franklin D. Roosevelt, who had taken a keen interest in them as a New York state senator, then as governor of New York, and then in the White House, when he established the federal CCC and conservation work was thus expanded to a wide range of federal and state forest, park, and soil conservation agencies. The forest industry was as enthusiastic about CCC work as the state and federal agencies were, and during debates leading to its demise the industry urged that the program be continued.

The states became one of the major vehicles through which close relationships between the industry and the Forest Service could be forged. The agency established the State and Private Forestry Division, which distributed grants-in-aid, and the programs they funded became crucial elements of most of the state forest agencies.[13] Over the years, additions to this earlier schedule of fire and replanting programs were made, and varied schemes of state regulation of private forest management were attempted. Those schemes were largely remnants of the more ambitious regulatory proposals of earlier years. All of this activity seemed to establish firmer relationships between industry and state forest agencies rather than the federal Forest Service.

The increasing rapprochement between the industry and the agency was exemplified by a new strategy: the Cooperative Sustained-Yield Program, authorized by an act of Congress in 1944.[14] This law anticipated joint agreements between a timber company and the agency in which their holdings would be managed as a single unit, with the company agreeing to comply with sustained-yield management and the agency agreeing to give the industry exclusive rights to cutting on the national forest lands. Despite glowing claims that this arrangement was the wave of the future, only one such agreement was realized, with the Simpson Lumber Company on the Olympic Peninsula. Other proposals fell by the wayside as local firms and residents complained that the proposed agreements anticipated an unfair monopoly sanctioned by the Forest Service. It also turned out that experience with the Simpson cooperative sustained-yield unit did not fulfill its promise, and the glowing predictions for the program soon turned dark.

An even more ambitious attempt at reciprocal cooperative relationships between the industry and the Forest Service took place in Alaska.[15] In the southeast panhandle of Alaska, President Theodore Roosevelt created the Tongass National Forest in two executive orders, in 1902 and 1907. From its inception, Tongass was viewed by the agency as an enormous opportunity to conduct a massive forest management program.

Tongass, the largest national forest in the country at 17 million acres, contained a huge amount of timber. However, there was no wood-consuming industry in the region, so the agency's strategy was to create one. After a number of unsuccessful attempts, the boom in the lumber market during World War II gave rise to a plan to negotiate fifty-year contracts with the Ketchikan Pulp Company and the Alaska Pulp Corporation to construct and operate mills at Ketchikan and Sitka.

These contracts were unique for the agency; they meant that clear-cut harvesting and road building would dominate the agency's objectives for the

Tongass National Forest. This management strategy remained in place when the Alaska Native Claims Settlement Act of 1971 created thirteen native corporations and allowed them to select more than 500,000 acres of Tongass, which were almost all clear-cut within twenty years.

By the late 1960s, these contracts had become the subject of considerable public debate; groups in both Alaska and the lower forty-eight demanded reforms in the program. In legislation passed in 1980, 5.4 million acres (of which 3.6 million consisted of rock, ice, muskeg wetlands, and noncommercial forest) were protected from development, but its provisions also set a timber target of 4.5 billion board feet a decade and established a permanent timber appropriation of at least $40 million a year to support the industry. In another "reform" act in 1990, the timber targets from the 1980 legislation were revoked, the $40 million annual appropriation was repealed, and modifications were made in the long-term pulp contracts. Thus, the agency's silvicultural ambitions once again faced the realities of the new forest economy that was gradually evolving in the middle of the twentieth century.

Wartime and postwar markets for wood established closer relationships between industry and the agency in their joint effort to supply the demand. The wartime levels of demand remained high even after the war because of the booming housing market. The agency allowed timber harvests to grow rapidly, so by the end of the 1950s, both industry and agency were accustomed to larger harvests, and they continued their close cooperation. From the industry perspective, this close cooperation was stimulated by the fact that its own supplies had dwindled and it was looking to the national forests to replace them; they pressed to increase the harvest even further.

From the agency's perspective, the larger harvest seemed equally attractive as a permanent fixture. The Forest Service rapidly translated its long-standing objective of sustained yield into a rationale that advanced maximum yield, based not so much on earlier theories of sustainability but on theories of just how much wood production could be generated.[16] To support this strategy, the agency developed the idea that with increasing use of genetically selected seed, fertilizer, herbicides to eliminate competing vegetation, and other "inputs" they could safely predict increasing production in the future.

With that in mind, they could safely harvest more in the present without damaging future supplies. This "allowable cut effect" provided the rationale for the increased harvests that both industry and agency desired.[17]

The agency-industry rapprochement did not last long, however. Increasing pressures from the public at large and the users that had come of age since 1920 called for scaling back rather than increasing the allowable cut in order to reserve timberland for objectives other than wood production. These demands were an integral part of the emerging environmental and ecological concerns being expressed by both the public and biological scientists. More agency resources were needed to satisfy those objectives. While the agency often tried to assure the public that it could satisfy these desires just by increasing its "output" for all, this did not in fact seem to be possible, producing what historian Paul Hirt has described as a "conspiracy of optimism."[18]

Faced with such pressures, the industry sought relief from Congress and the executive branch, asking the government to curb the "diversion" of agency resources from wood production.[19] The Forest Service could not seem to move far from its wood production roots, its personnel having been trained in that management culture for many decades; it thus tended to share the industry's belief that the main objective of the agency was to grow and harvest trees.[20]

Industrial Forestry

Along with this rapprochement in forest policy issues, the Forest Service and the industry tended to come together on a number of technological developments in wood production that marked the growth of "industrial forestry."[21] This surge involved, first, advances in the science and technology of tree growth to increase the wood fiber in each tree and, second, changes in forest technology and forest management aimed at reducing labor costs. The Forest Service devoted considerable time and resources to fostering these "progressive" developments and in so doing forged closer ties with industry in a common effort to increase wood production. At the same time, this agency preference continued to subordinate other multiple-use objectives, such as wildlife, watershed protection, and aesthetics, to this dominant focus.

The technology improvement and labor reduction efforts that began in the 1920s and 1930s and grew steadily came to the fore in the 1950s around the objective of increasing the nation's wood production, a vision stimulated by the accelerated wood harvest of the World War II years. Both technology and labor innovations reflected changes in the circumstances of the wood production industry, which prompted it to practice continuous management rather than the previous tendency to harvest and then leave the task of "growing a new crop" to the Forest Service.[22]

As this change in the industry's attitude toward continuous wood production emerged, wood producers tended to think in terms of a division of labor in which they would engage in production and the agency would carry out the research to enable them to do so. Clepper cites the proceedings of a national conference on commercial forestry sponsored by the U.S. Chamber of Commerce in Chicago in 1927 at which D. C. Everett, president of the American Pulp and Paper Association, emphasized the need to educate the public so that it would support larger appropriations for research. Research was said to be the key to all the industry's ills, and industry believed it was the responsibility of government to conduct experiments in management and utilization for industry.

One aspect of this effort to apply forest technology involved attempts to grow trees that were more desirable as raw material and to make them grow faster. Pressures of this type were fostered especially by the growing pulp and paper industry that sought to reduce the age at which trees could be harvested for pulp. One experimental step involved the development of seeds that were more productive, an effort that had been under way for many decades but that had produced useful results only just after World War II.[23] Having produced genetically better seed, researchers then applied such practices as fertilization, insect and pest control, herbicides, and fire prevention to improve yield. All of this work involved parallel efforts on the part of both industry and the agency, and it further enhanced the role of wood production in national forest management.

Equally significant in this agency-industry relationship were efforts to reduce labor costs by shifting the focus of management from individual trees to large forest areas. To the public the most dramatic aspect of this shift in management focus was the "clear cut," the practice in which trees were harvested not by selecting individual trees but by harvesting all trees in a given area uniformly, much as one would harvest grain. Innovations in technology that reduced the labor required to harvest a given area were driven by industry efforts to lower costs; they were also taken up by the Forest Service.

Clear-cutting, which is highly visible, was only one aspect of a shift toward area management rather than single tree management. Inventories of growing stock, once determined by individuals who went "timber cruising" in the woods, were increasingly replaced by aerial photographs from which timber volumes were estimated. Tree growth in the new forest was tracked and monitored by "even-age" management, which reorganized the forest from one of trees of mixed ages to one of trees with uniform ages. The entire forest was measured, followed, shaped, and harvested as an area rather than as trees of varied species and ages, all of which could be followed closely through modern computerized information, replacing the earlier, more varied systems of management and reducing labor costs for almost every aspect of wood production.

Most of these changes took place slowly and unobtrusively, and their implications for management objectives that sought to replace the patterns of more natural forest processes were slow to be widely known. But harvesting by area management principles was quite visible. Large numbers of Americans who had various types of contact with forest affairs could see, with their own eyes or in media images, the results of such harvesting, and they were appalled. The agency and the industry sought to hide their clearcuts by maintaining "beauty strips" between cuts and roadways or hiking trails, but to no avail. The cuts were quite visible to hikers from the ground and from aerial images that were widely circulated.

Legal action arose in Montana and West Virginia on the grounds that the Forest Management Act of 1897 allowed only "mature" trees to be cut and required that they be marked for cutting. Clear-cutting violated both provisions, the plaintiffs argued, and a federal court agreed, deciding the

Monongahela case in favor of the plaintiffs.[24] The Forest Service agreed that clear-cutting was not authorized by law, but its defense was simply that forestry had changed since 1897. And well it had. The technology of forest management, evolving steadily since the early 1920s, had come to fruition after World War II, spearheaded by industry's search for lower costs and influencing both public and private management. The implications of these technology advances for the future of the U.S. Forest Service were far reaching.

The major significance of the growth of industrial forestry was that its tenets called for imposing a simplified management regimen on a highly complex ecological forest system. As the ecological forest became more visible in the years after 1970 and the public demanded that managers recognize forest ecology more fully, it became increasingly apparent that industrial forestry was ill prepared to cope with this new world of forest affairs. The limitations of this management philosophy were responsible for many of the emerging conflicts over forest policy.[25]

Grazing Management

In managing its grazing lands, a task the Forest Service had inherited at its organizational birth, the agency found itself in much the same position as with wood production. An initial emphasis on the deterioration of the range through overgrazing and an ambition to restore the range by reducing its use was gradually influenced by market realities that resulted in paring down those initial objectives.[26]

The market circumstances of wood production and grazing differed. In the first case the agency was a supplier of raw material and found that it had to foster the creation of mills to create a market for timber from the national forests. In grazing, the agency was a regulator of private production, seeking to control use of the range by imposing requirements for stocking levels; it was thus seeking to protect rather than market a resource. In the case of wood production, the relationships between agency and user were focused sharply on the timber contract and, in the case of grazing, on the terms of the grazing permit. In both cases, an initial ambition to improve resource use with an eye toward "conservation" objectives was modified considerably as the relationships between the agency and its clientele evolved over the years.

Two economic realities influenced the grazing program. The first was that the agency's grazing officials knew as little about how to determine the condition of the range and an acceptable number of cattle to include in the permits as did the stock owners themselves. Agency personnel thus thought of the range's ecological health in economic terms and were heavily influenced by the prosperity of the livestock industry. Trends in carrying capacity goals usually followed economic conditions, rising with good times and falling with bad times.[27]

The second reality was that the economic circumstances of both the livestock industry and the federal grazing program tended to give the grazing permits an economic rather than an ecological value.[28] In other words, while to the agency the permit represented an acceptable "carrying capacity," to the permittee it represented economic value. The permits could not be bought or sold, but in the market for loans, the permits were economically valuable to the livestock owner since banks included them in determining the economic viability of the loan. As a result, the larger market circumstances tended to define the permit in terms of its significant economic value, and the agency was thus caught in the economic web that tended to shift the meaning of the permit from an instrument for improving the range to one of economic value in the market.

By 1920, the permit system was well established, as were the issues that tended to shape the relationships between the agency and stock owners. A focal point of these relationships was the array of stockmen's associations that the Forest Service had encouraged by giving them responsibility for allocating livestock among the potential permittees.[29] The activities of these associations, however, soon went far beyond that initial task, and they became important voices for the industry in its negotiations with the agency on a variety of issues. The most critical of these were the issues of redistribution of permits to new applicants, the duration of the permit, and the demand that the permits be recognized as a property right, subject to sale or transfer.

The agency weathered these disputes rather well. At first it sought to "democratize" the program by awarding allotments to stockmen with smaller operations; this move met opposition from the stock industry. However, the agency itself considered the larger owners more amenable to measures to protect the range, more reliable, and easier to work with, so it phased out its "redistribution" policy.[30]

The second grazing problem was more serious: the ongoing attempts by permit holders to retain their grazing privileges, first by demanding long-term (e.g., ten-year) permits and then, in more extreme cases, demanding that the "privilege" of receiving a permit be translated into a legal right to the range that could be bought and sold.[31] As to the demand for ten-year permits, the agency gave way to the industry in 1924 and agreed to it, but a decade later in 1934, it declined to renew the policy. The agency was most concerned about the demand for a property right in the permit and was able to fend off that continuing proposal. In all of these issues the industry received considerable support from some key western senators, but the Forest Service came out of the tussles relatively unscathed.

However, it was not able to reduce stocking levels effectively. Debates persisted over the issue of whether or not the forage was improving or livestock numbers should be adjusted downward. The agency's ambitions on that score were continually thwarted by the economic circumstances of the industry it regulated.

Grazing in the eastern national forests that had been acquired from private owners took an entirely different turn.[32] These were cut-over lands, many of them at the early stages of regrowth into "second growth" forests at the time of acquisition. Over the years these forest lands had been used by farmers who let their livestock (cattle and hogs) graze on the forest lands that were regenerating with new trees. This type of grazing posed a problem for successful forest regrowth since the grazing animals ate the emerging seedlings as part of their forage. As a result the agency simply prohibited the use of the forests for grazing as the solution to their problem. The reduction that was accomplished only through years of painful grazing management in the West was solved almost by a stroke of the pen in the East.

Accommodating Outdoor Recreation

Recreation was a rather new use of the national forests with which the agency had to cope.³³ It had not been an active element in the agency's affairs prior to World War I, and only after that war did its long and permanent role in the national forests begin to make itself felt. At first, its impact came in two stages. One involved an agency initiative, when the Forest Service took an active interest in the construction of summer homes and resorts and the activity of professional recreation guides in the forests. The other stage involved the presence of an increasing number of campers, whom the agency found it difficult to avoid even though it often indicated that it wished to.

Regarding the former case, in which recreation was an active agency issue, historians have not taken much interest, and consequently only the bare outlines can be described at this point of historical inquiry. The Term Permit Act of 1915 is a starting point.³⁴ That law authorized the Forest Service to grant permits for homesites, resort sites, and similar recreation-oriented purposes. The source of the law might well be thought of as those who sought to initiate such developments and therefore urged the agency to propose such a measure. But the agency was equally interested, since it was under considerable pressure to "make the national forests pay" and was having some difficulty in making good on its prediction that it could do so. At any rate, with the authority of the 1915 act, the Forest Service began to identify possible development sites and instructed the district foresters in the West to begin such an inventory. Aldo Leopold, in the Southwest's District 3, took on this inventory as one of his responsibilities, and Arthur Carhart, a landscape architect, was employed in March 1919 at the Denver office in District 2 to do the same.³⁵

Carhart's new responsibilities were far more extensive than this limited task, however, for District 2 seemed to be especially primed for outdoor recreation. Early in 1917, two years before the agency hired Carhart, a meeting of the region's forest supervisors had given much attention to recreation possibilities. In a talk given at that meeting, Wallace Hutchinson, in charge of the district's information and education branch, had urged the agency's staff to "open up" the national forests for public recreation and "make the mountain forests of the west the favorite vacation lands of people who take a summer outing."[36]

Carhart thus assumed responsibility for a relatively unformulated but potentially broad-based outdoor recreation program. He worked especially with the Pueblo, Colorado, chamber of commerce near the San Isabel National Forest to foster recreational activities there, but he became more widely known for his approach to using Trappers Lake in Colorado for forest homesites. He argued that the forested lands around the lake in which homesites were likely to be established should not be logged in order to maintain their "natural beauty" and thus their appeal to potential forest residents.[37]

Carhart explored a range of recreational activities and drew up a number of extensive plans for recreation in his district. He took a special interest in recreational opportunities in the Superior National Forest in the Boundary Waters Canoe Area and in particular its potential for a wilderness-type canoeing experience. But he found that his efforts were not appreciated by the agency's personnel in the field nor by the Washington office when it came time to translate plans into budget. Amid quiet indifference from the agency personnel—Carhart was known, with some derision, as the "beauty engineer"—and amid the failure of the national office to support his program, he resigned his position in Denver and thereafter was known primarily as an outdoor writer and enthusiast for outdoor recreation in general.[38]

It seems clear that Forest Service personnel had varying opinions about recreation. In 1917, it appointed a well-known landscape architect, Franklin Waugh, to investigate the western forests for their recreational potential. In his report to the agency, *Recreation Use on the National Forests,* Waugh recommended that the Forest Service hire men "suitably trained and experienced in recreation, landscape engineering, and related topics."[39]

The key problem was the lack of professional recreational leadership in the Forest Service. Men trained in "technical forestry" were in charge of silviculture and experts in the cattle industry were in control of grazing; recreation required the talents of the "landscape engineer." But it was precisely because of the agency's refusal to modify budgets to provide for such leadership that Carhart had resigned.[40]

The role of recreation in the national forests soon took quite a different turn. Initiatives for addressing the recreation objective came less from the agency and more from the general public, which had imposed its presence upon the forests and which the agency looked upon as a problem and a burden with which it had to cope in a defensive mode of action. Furthermore, early public interest in nature had been expressed primarily by the well-to-do, who could easily afford the cost of transportation to wilderness areas and of food and shelter once there. With the advent of the automobile and all-weather roads increasing access to forest areas, Americans from other ranks of society were rapidly taking an interest in the outdoors. Although the national parks were among the most attractive destination points, the national forest lands were far more extensive and more accessible.

With hardly any effort on the agency's part, the national forests became a magnet for "car camping."[41] Outdoor recreation, which had begun in the early 1920s, accelerated with each passing year, even in the Depression, with only a brief slowdown during World War II, and the influx of campers overwhelmed the forests.[42]

In its annual budget proceedings, the Forest Service continued to argue for additional funds to handle this increasing demand for recreation. The implications were inescapable for the agency. Historian Paul Hirt reports that at a Society of American Foresters meeting in 1947, an agency forester participating in a panel discussion put the agency's dilemma succinctly: "There is no point in trying to explain this recreational urge of our people. Its existence and its imperious demands are demonstrated facts which we cannot ignore."[43] But, says Hirt, "most professional foresters had nonetheless resisted accepting recreation as a co-equal partner with timber production." Amid increasing conflicts over recreation, "the agency tried to contain these battles and keep decisions from being carried over their heads, but to no avail."[44]

Unlike the home and resort occupants, campers were not an economic asset; they provided the agency with little income. Moreover, because the campers could pick and choose their camping sites, they could easily interfere with the agency's "normal" activities. In addition, they built their own camp fires and often were the cause of forest fires, the bane of agency administration. The Forest Service responded in a defensive mode by adopting a zoning strategy. Camping was permitted only in areas authorized and developed for that purpose. Acceptable tent sites and related parking areas were prepared and identified. Moreover, camp fires could be set only in specific fireplace units. Thus, a zoning system brought this recreational use of the forest under control. A system of selected lands used for selected purposes was inaugurated and became a model for similar problems in the future.

Outdoor recreation received a considerable boost from the Civilian Conservation Corps's construction of facilities in the 1930s. These construction projects were carried out by the CCC on recreational lands in general, parks as well as forests, and on state as well as federal lands. The Forest Service benefited considerably from CCC-built campgrounds and associated recreational facilities such as hiking trails and swimming areas, all of which expanded facilities that were already becoming extremely popular.

By the 1960s, several of these recreational construction projects, especially the hiking trails and the canoeing facilities, had evolved into national programs fostered by two laws in 1968: the National Trails Act and the National Wild and Scenic Rivers Act.[45] Both laws applied to all federal public lands, and both involved the zoning strategy first adopted by the U.S. Forest Service for car camping. These two acts, however, zoned not only the areas of direct recreational activity but also their surrounding environment—the visual buffer zones that implied the close association of recreational activities and aesthetic values, values that the agency felt compelled to work into its management strategy in spite of its long-standing impulsive rejection of aesthetics as an authorized responsibility.[46]

Outdoor recreation gave rise to an outdoor recreation industry; one grew with the other, so much so that the agency found itself faced with a new and active economic interest to which it began accommodating itself because of the potential financial benefits.[47] With car camping came the accouterments supplied by the camping industry—the gasoline-fueled cook stoves, the folding tables, the cooking pots and eating utensils, the tents and sleeping bags. At the start of the boom, campers merely brought items from home or surplus army equipment, but gradually a consumer industry began to make versions of the same equipment more appropriate for outdoor use.

As recreation moved beyond car camping to such activities as hiking, backpacking, and canoeing, each activity gave rise to new products and types of equipment, such as light-weight packs, dried food that could be reconstituted, light-weight tents and bedrolls, and canoes made of light-weight metal and fiberglass. The outdoor equipment business became one of those economic interests that the Forest Service found it had to "bargain" with in its management program.

Outdoor recreation also came to be something like a competitive sport between the Forest Service and the National Park Service.[48] Both agencies were subject to the growing popularity of outdoor activities, but the Park Service looked upon the agency as an unwarranted competitor. The Forest Service was very concerned about this view because there was a tendency for wilderness advocates to seek to carve a new national park out of a national forest. Over the years, the forest agency had been losing land to the parks agency.[49]

The Forest Service had long noted the tendency of nature enthusiasts to propose a national park for particular areas because the National Park Service readily accepted aesthetic forest objectives. This had led the Forest Service to make enough accommodations to aesthetic objectives that the two agencies continued their competition, at times quietly but at other times openly and vigorously. Historians have said, with some justification, that after World War II, when the Park Service inaugurated its expanded recreational program known as "Mission 66," the Forest Service soon responded with its "Operation Outdoors," a similar attempt to upgrade its outdoor recreation facilities.

The Forest Service and the Park Service had quite different responses to the recreationists with whom they now had close contact. The Park Service looked upon the presence of these visitors as an educational opportunity; it considered the park setting one in which it could encourage larger numbers of the American public to appreciate and study nature.[50] Park officials promoted use of their lands as opportunities for the study of natural history, and many a park developed its circle of both specialists and amateurs who produced publications about their park's nature resources. This educational opportunity was extended to a larger audience of park visitors through both camp-fire and field trip interpretive programs, all to establish a sort of bond between park visitors and staff that enabled the national parks to firmly root themselves in the American public's awareness. Moreover, these activities fostered a pattern of institutional commitment that enabled the Park Service to serve the new interest in the ecology of the public lands that later emerged.

The Forest Service, on the other hand, tended to view its recreational facilities not as an opportunity for nature appreciation and education but as a defensive management strategy to control and contain a public presence that was not relished.[51] What limited outreach to the public that it did undertake, moreover, had a strong tinge of commodity forestry, emphasizing the high rank of wood production in its activities, thus precluding a more comprehensive and balanced understanding of the forest as a reservoir of ecological resources and understanding. The Forest Service did not enjoy the firmer bond between visitors and staff that the Park Service developed.

Wildlife Management

Wildlife had long played an important role in the minds of those interested in the forest reserves; the Boone and Crockett Club, the group of wealthy sport hunters, had taken the lead in supporting the reserves during both the run-up to the 1891 law proclaiming the reserves and the subsequent development of their management program.[52] Pinchot, however, had taken a dim view of their inclusion in the agency's objectives and had rejected the club's main proposal for wildlife in the national forests, namely that the forests serve also as wildlife refuges.[53]

In the years after the transfer of the reserves, however, interest in sport hunting grew so rapidly that in the 1920s the number of hunters was more than double that of the previous decade, reaching an estimated 6 million.[54] Much as with camping, automobile access had played a crucial role in interesting a wider swath of American society hunting wildlife in the national forests. This group, so much larger in numbers than the old Boone and

Crockett Club membership, reflected a new stage in the involvement of the U.S. Forest Service with wildlife.[55] The Izaak Walton League, a midwestern group of outdoorsmen organized in 1923 with a Chicago headquarters, now became one of the most active advocates of public hunting and fishing habitat, and its members were a critical influence in the protection of the Boundary Waters Canoe Area in northern Minnesota.[56]

Over the years since the mid-nineteenth century, hunting and fishing advocates had prompted the creation of a complex of laws and administration that emphasized state-level wildlife programs; the foundational principle was that wildlife was a public resource and subject to state rather than federal ownership and management.[57] Therefore, one of the main considerations in potential wildlife management was the Forest Service's relationship to the state programs.

Those programs had, for the most part, emphasized controlled hunting —restrictions on the "take" of fish and wildlife so as to arrest declines in wildlife populations. Central to these programs was enforcement of the restrictions, and a system of fish and game wardens had evolved to implement that enforcement activity. But this enforcement proceeded slowly because of the lack of state personnel. What role would the Forest Service play in this enforcement activity? The nation's most significant populations of wild game were in the national forests, but for many years the agency took no active steps to deal with them.

Aldo Leopold, who was associated with the Forest Service in District 3 in New Mexico, took up the enforcement of state restrictions as a major interest, and the district supervisor encouraged him in that effort.[58] Leopold helped to organize the New Mexico Game Protective Association, was active in New Mexico state game affairs, and sought without success to interest the Washington office of the Forest Service in wildlife. One of Leopold's major suggestions to agency officials was that forest rangers should assist state wildlife agencies in enforcing their game laws and be compensated for time spent doing so.[59] The Washington office, in the person of Leon Kneipp, assistant chief forester, responded that game protection should not be considered as part of Forest Service personnel's jobs but as a "matter of public duty." If not, then "we should disavow the responsibility for the game and refuse to be connected with its protection."[60]

During the years after World War I, what little in the way of resources the Forest Service devoted to wildlife was technically a subordinate part of its grazing program. In fact, its meager staff of wildlife personnel was devoted to objectives desired by the livestock owners, and until 1937, such staff were employed in the Division of Grazing.[61]

Their work concerned two major policies. One involved the eradication of predators, which the stockmen blamed for significant livestock losses. Agency personnel joined in the antipredator work. The other involved competition between wildlife and livestock for public land forage. This was a more difficult situation for the Forest Service, for it meant choosing between the needs of livestock, grazing being a long-established use of the range, and the needs of wildlife, a still growing use of the range. Managing these two areas of concern would inevitably involve making choices to advance the use by one and restrict the use by another.[62]

The work of local livestock committees told the tale of just how these competing interests worked out. The Forest Service had established those committees so that they, rather than the agency, would decide what livestock would use the range. They were dominated by the livestock owners, and among ten or twelve members there was usually only one representative of a game organization.[63]

After World War I, the role of the Forest Service in wildlife affairs took quite a different turn, shifting from establishing refuges to promoting increases in game to protecting forest vegetation from the increased numbers of browsing animals. This shift was highlighted by the problem in the Kaibab Forest Game Reserve in the Southwest, where large numbers of deer destroyed forest vegetation. The agency thus began to seek ways to reduce game population rather than increase it. This produced conflict between the Forest Service and sportsmen and their allies in state game agencies, who denied that game was too abundant and opposed Forest Service policies to reduce it. For several decades the Forest Service and sportsmen's organizations were at odds over the potential damage to forest vegetation caused by deer.

This shift in the context of forest wildlife also led to a divide between the states and the Forest Service, a divide that followed the lines of two Supreme Court cases: the *Geer v. Connecticut* decision in 1896, which placed responsibility for wildlife in the states, and the *Hunt v. United States* decision in 1928, in which the Forest Service was authorized to protect forest resources from deer overpopulation. Just where one drew the line between these two authorities on specific issues remained controversial and was often a matter of different accounting methods used to come up with game population numbers. Not until late in the 1930s did some agreement emerge on that issue, as professionals on both sides began to establish common ground on how to determine game population levels. Far more significant for the long run was the way in which these problems fostered an increasing emphasis on habitat as a central factor in national forest wildlife management.[64]

During the 1930s, there was a marked increase in the public's interest in wildlife. Some of this increase came from the initiative of President Franklin D. Roosevelt and some from the initiative of Jay "Ding" Darling, who took up leadership in wildlife affairs within the federal government. Several organizations promoting wildlife interests also arose during this period. The National Wildlife Federation, an association of state game organizations, was formed in 1937. The Wildlife Society, an organization of wildlife professionals, was also formed that year, as was the Wildlife Management Institute, composed of wildlife leaders both government and private, which promoted national wildlife policy.[65]

Amid all this public interest, the Forest Service took tentative steps toward a more expansive wildlife program. In 1936, it established its Division of Wildlife, separate from the Division of Grazing, with eighty-three game specialists. But this administrative scope was curtailed during World War II, and after the war, the agency reverted to its prewar initiatives only slowly.[66]

In taking these steps, however, the agency had some hurdles to overcome. In a forest that was in the early, youthful years of growth, that is, in an "early successional forest," wildlife and wood production objectives could easily be integrated, for the young forest provided food for browsing animals such as deer and grouse. But some game, such as the bobwhite quail in the

South, thrived on the undergrowth that usually was abundant after repeated fires, which of course hampered the growth of trees for lumber. One professional who promoted habitat for quail, Herbert Stoddard, found himself at odds with foresters on the issue of what came to be known as "fire ecology," which the foresters strongly opposed.

A growing interest in wildlife on the part of foresters was reflected in the role of the Society of American Foresters, which issued two reports on the relationship between forestry and wildlife, both of which were published in the *Journal of Forestry*. Both described the difficulties that commodity forest advocates had as they attempted to absorb wildlife objectives into their management strategies. One report, published in 1937, was titled, "Report of the Committee on Game Management with Reference to Forestry"; the other, "Second Report of Game Policy Committee," appeared two years later. This second report described the problem of integrating forestry and wildlife: "The difficulty is in large measure due to the lack of appreciation of the real meaning of wildlife management. In cases where those in charge of a forest think that the information incidentally picked up while fishing or marking timber in a watershed is all that is needed for management of its wildlife, we cannot hope for much progress."[67]

In 1944, Lloyd Swift, an agency employee with a reputation for a strong interest in wildlife, became chief of the Division of Wildlife, yet his salary was paid for with discretionary funds. Even this support was precarious. In 1947, Congress eliminated the wildlife budget for the 1948 fiscal year, and it did so again for 1949. Not only Congress but also the agency displayed this lack of support. In an oral history interview some years later, Swift explained in some detail the circumstances for wildlife within the Forest Service. He emphasized that "up until the time that the Division of Wildlife Management was created in 1936, wildlife was part of range management and ... even by 1968 the wildlife employees in the western regions were in the Division of Range Management." He deplored this situation on the grounds that both timber and range people viewed the forest narrowly: "Very often we have timber people who see nothing but timber, range people who see nothing but range." He indicated, however, that "that's not characteristic of wildlife people. They're pretty sensitive to coordinated land use."[68]

During the years following World War II, the evolution of wildlife affairs took several new and quite different turns. While game and game hunting had dominated throughout the first half of the twentieth century, a succession of innovations relating to nongame species, then endangered species, and finally biodiversity and the entire range of "wild" forest species took place. All of these came to be expressed through a wide range of public activities and scientists.[69]

For many years the Forest Service had limited its vision of wildlife affairs and had become relatively indifferent toward them. Thus, by the 1970s, when wildlife resource issues were assuming an ever-larger role in public land affairs, the agency had neither the experience nor the expertise to cope with emerging wildlife objectives. Over the years, the agency had gradually embraced recreation and developed considerable expertise in coping with it. But the agency's role in wildlife affairs continued to display a marked absence of leadership and a lack of agency competence, which it inherited from decisions made at the agency's very beginning to give wildlife only limited attention.

Watershed Management

From its earliest days the Forest Service had continued to follow the statutory mandate of the 1897 law that one of the two main objectives of the reserves was watershed protection. But, as has been observed earlier in this book, it hesitated to move forward vigorously with an appropriate management program. This limited approach seemed to be closely related to its caution about blaming sheep grazing for headwater erosion, which was the main focus of western watershed concerns at the time. This caution was expressed in the decision by the agency to investigate these "forest influences" considerably further before taking any action.

A more vigorous approach to watershed protection and erosion was suggested by Aldo Leopold in District 3, which encompassed New Mexico and Arizona.[70] Leopold's responsibilities took him around the forests of the region to observe the results of agency policies, and in the process he developed a keen interest in the condition of watersheds and made a systematic attempt to understand the role of forests in soil erosion.

Leopold argued that the problem was most acute in the mountain valleys, which supported the agricultural activities that supplied the lumber and mining towns. He recorded that seven of thirteen such agricultural valleys in Arizona with which he was personally familiar were in danger of deterioration through stream bank erosion.[71] This problem was due, he argued, to the concentration of animals in livestock service areas such as watering places, salt grounds, drive ways, shearing pens, and roundup grounds.

Leopold's analysis of the issue brought a new perspective to the causes of watershed problems. While the agency traditionally had been looking at the control of livestock in the headwaters as the way to protect watersheds, Leopold focused on the pressures of livestock use in the downstream riparian areas.[72] Due to overgrazing and overuse there, the sod was weakened, stock trails had started gullies, and brush stabilizing the creekbed had been destroyed. From talking with old-timers, Leopold concluded that the damage started when the range began to be used for stock. Grazing had ultimately weakened the resistance of river banks to erosion from floods. All of this new information demanded a shift in perspective from headwaters management to managing grazing in the riparian valleys.

Leopold's views about watershed/erosion management were shaped in part by his observations of the Prescott National Forest in Arizona, which had been established originally as a watershed forest, that is, to protect the watershed values of the forest and not the timber itself.[73] He recommended, "We should restore the grass which all the evidence indicates is better watershed cover than either brush or woodland," a proposal that Leopold biographer Curt Meine describes as a "revolutionary recommendation for a Forest Service employee."[74] The specifics of Leopold's recommendations were to "maintain grass cover on the watersheds, especially on the watercourses; try to keep and restore the willows along the stream banks; use inexpensive artificial works to clog channels; work with other landowners on a unified plan."[75]

Leopold compiled a record of his investigative experiences and subsequent recommendations into a watershed handbook, which he completed in 1924. He intended it to serve as a guide to teach personnel how to diagnose and respond to watershed problems.[76] In the handbook he criticized agency range policy, which had long maintained that a given range could accommodate more livestock as long as the forage was there and as long as grazing helped to reduce the fire hazard. Leopold argued that the number of livestock under permit should be determined by the overall condition of the watershed itself. It was a matter of managing the forest and range resources rather than managing the cattle.

Historians generally have neglected to examine the watershed and erosion aspects of agency responsibilities. But in the absence of more detailed evidence one is prone to conclude that Leopold's recommended approach did not take very firm hold. He summarized his judgment about the results of agency management on the Tonto National Forest, which was within the range of his activities in District 3, with this observation: "the timber resource on the Tonto has undergone a vast improvement, while watershed and range values have undergone a vast deterioration."[77]

The idea of streambed erosion as a component of watershed management, which Leopold emphasized, had few advocates within the agency and seemed to languish. The possibility that watershed issues would come to the fore depended far more on the degree to which those affected by forest practices might speak for water users. One such case involving drinking water appears early in the agency's history, when towns and cities expressed concern that forest harvest practices might harm their drinking water. Few historical accounts focus on this problem, but several pieces of historical evidence suggest the need to pursue it further. An agreement between Grand Junction, Colorado, and the district forester in Denver may illustrate a scenario in which watershed issues arose because of water consumers.[78] In this agreement, signed in 1915, the town's concern for its watershed on national forest land was acknowledged, but the Forest Service retained the right to carry out its forest practices as it saw fit, with assurances that it would not harm the town's water supplies. In later years, in other locations, it was just that assurance that towns called into question, leading to conflicts between agency and drinking water consumers.

Several instances occurred on the Olympia Peninsula in which the Forest Service, despite local protests, proposed to log urban watersheds. One controversy, involving the city of Port Angeles, Washington, on the peninsula's north coast, was resolved with action by President Franklin Roosevelt, who added the area to the Olympic National Park.[79] Similar controversies were reflected by the introduction of the Municipal Watershed Act of 1940, legislation the agency supported because it was intended to restrain possible legal action by municipalities in similar instances. The "law discouraged cities like Port Angeles from objecting to logging in their designated municipal watersheds by making them liable for the 'losses' to the federal Treasury of any net revenue that might have been gained from the forgone timber sales."[80] The law was never used, presumably because its potential financial penalties deterred action by municipalities to protect their watersheds from logging, but according to historian Paul Hirt, it is still on the books.

Historian David Clary reports another case that arose from proposed logging on the Olympic Peninsula. When the agency proposed logging on the Big Quilcene watershed of the Olympic National Forest in 1955, the municipality of Port Townsend expressed concern about the possible effects on its water supply. In reply, the Forest Service assured the town's officials that it would "propose nothing ... that will adversely affect the amount and purity of the water supply." The agency explained that it was necessary to carry out the cut, concluding, "It should be stressed that timber harvesting on the watershed is inevitable and to delay longer means more extensive and severe cutting when the completion of the rotation in the Working Circle eventually forces logging activity onto the watershed."[81]

Paul Hirt cites the multiple-use plan for the Canyon Creek Ranger District of the Gifford Pinchot National Forest as further evidence of the agency's tendency to minimize watershed problems arising from timber harvesting.[82] Agency regulations emphasized that areas with unstable soils that could not effectively be protected from erosion during logging should be excluded from the commercial timberland base. However, as allowable cut demands increased, foresters and engineers adjusted the level of erosion that would be tolerated and adopted optimistic assumptions about their ability to engineer safe roads on almost any soil type. They worked under the philosophy that proper management and use of technology could overcome the possibility of environmental harm.

The watershed responsibilities of the Forest Service took on new dimensions in later years as those who faced potential harm from forest watershed activities tended to redefine the relevant issues. Groups in that category included those interested in the fishery or in drinking water and agencies reflecting those interests, such as the U.S. Fish and Wildlife Service, tribal fishing interests, the Environmental Protection Agency, and advocates of endangered species protection. But the agency continued to minimize watershed problems, just as it had in the five decades after World War I, when a pattern of relative indifference to its watershed responsibilities, decreed in the 1897 act, was much in evidence.[83]

Parks and Landscape Aesthetics

Aesthetic objectives were as difficult for the Forest Service to cope with as wildlife objectives. The public had expressed strong interest in both since the earliest days, and this persistent interest had had an impact on forest affairs ever since the 1891 law authorizing the reserves. In the early twentieth century, national parks had been the main proposal around which the aesthetics issue swirled, and because the Forest Service could not see itself administering parks with a "no timber cutting" policy, those interested in aesthetic forest values had rallied behind the proposal for a separate national park administration.

While this development led to dramatic agency controversies, a more mundane and even more extensive development of aesthetic policies emerged from the increasing vehicle travel in the national forests. The agency now faced increasing public revulsion against wood harvest practices that they could see for themselves while traveling for pleasure in the forests. Roadside areas that had been harvested were left with the unsightly remains of logging debris. This situation led to considerable debate within the agency. Could logging proceed to the edge of the road without adverse public reaction or should the roads be lined with unlogged strips?[84]

Some argued that potential public reaction could easily be prevented by explaining that tree harvest was silviculturally desirable, but others argued that this would not overcome the adverse reaction. They bolstered their arguments by pointing to the role of forested "beauty strips" in the management of the Cornell University tract under the direction of Bernhard Fernow. In that agency debate, the aesthetic concerns had prevailed, and "travel corridors" became major sites zoned against clear-cut harvests.

In later years, the same aesthetic values were applied to both hiking trails and "wild and scenic" rivers, and often the aesthetic protection went far beyond the nearby strips of land to include more extensive "viewsheds." With time, in fact, the agency employed landscape architects, conducted landscape research at forest experiment stations, and brought aesthetic objectives to bear on the size and shape of clear-cuts.

The aesthetic issue was most prominent in these years in the rivalry between the Forest Service and the National Park Service.[85] The alternative values involved, wood production and aesthetics, which Pinchot had considered administratively irreconcilable, now became well publicized in the agencies' rivalry. Proposals continued to call for carving national parks out of the national forests, and most of these proposals came from those who lived within the forests and found the aesthetic values desirable for their homes or for their resorts or other recreational businesses. Some proposals came from campaigns on the part of national organizations such as the Sierra Club, longtime champions of the national parks.

The Shenandoah and the Great Smoky Mountains national parks in the southern Appalachians had been integral parts of the drive to create national forests in the East. Their advocates promoted them as national parks and enlisted the support of private benefactors as well as state governments that foresaw the economic benefits of tourism. These proposals moved ahead steadily in the 1920s.[86] More traditional in their administrative origins were the Olympic National Park and the North Cascades National Park, both of which arose out of the Forest Service's refusal to forgo timber cutting, which then led to action to create a national park.[87]

The National Wild and Scenic Rivers System, established in the late 1960s, expanded the agency's concern for the reaction of hikers, backpackers, and river recreationists to the environment of their outdoor activities. The program also extended the agency's aesthetic corridor policies (established for highways earlier in the century) to a wider range of recreational environments and in some cases extended the aesthetic areas from the immediate travel borders to viewsheds, often to the tops of nearby hills. The agency considered the public objection to clear-cutting to be largely grounded in aesthetics and sought to modify the size and outline of clear-cuts to make them less offensive.

In these aesthetic activities, the agency often employed landscape architects, much as it had initially hired Arthur Carhart, its first such aesthetic expert, and it financed research at some of its experiment stations to assess public desires and how they could be met. As a result, the agency identified a land-use category, landscape management areas, "where exceptional scenic and recreational values" required sensitivity on the part of managers. This usually meant only smaller clear-cuts, more selection cutting as opposed to clear-cutting, and more "beauty strips" in areas designated for such management. Apparently the agency expected to continue its timber production even in scenic areas by merely adjusting logging activities to make them less aesthetically objectionable. As Paul Hirt noted, the agency "fundamentally was unwilling to forgo development and fought tooth and nail to keep silvicultural options open."[88]

Wilderness Protection

In these years between 1920 and 1975, the most widely debated aesthetic policy for the national forests was wilderness protection.[89] This issue continued the debates about national parks but in another administrative setting.

It involved not just aesthetic forest values on the immediate borders of recreational activities and timber harvest areas but aesthetic values in much larger areas. As in the case of parks, the issue threatened to remove some areas from the agency's silvicultural management ambitions.

The history of the wilderness issue cannot be understood simply by the conventional mode of contrasting ideologies of "preservation" and "use." One can, however, trace its evolution and elaboration over the years as an issue about how specific forest land should be managed. Proponents of wilderness areas wished to keep large, intact, and unfragmented forest areas free of roads, human-made structures, and timber harvesting. To some, such as Aldo Leopold, the specific area he proposed as wilderness in the Gila National Forest in New Mexico was considered the perfect place for hunting wild game in a roadless experience. At one point Leopold advocated that the area be called the "Gila National Hunting Ground."

To Arthur Carhart, who proposed that the canoe country of the Superior National Forest in northern Minnesota be declared roadless, the purpose of such an area was to provide a full range of outdoor recreation activities in a natural setting. On the other hand, Bob Marshall, a well-known wilderness advocate of the 1930s, was a hiker and mountain climber whose proposals sought to establish a forest setting for the physical and spiritual benefits of mountaineering.[90]

As proposals to identify and reserve as wilderness tracts in the national forests arose and were debated among national forest leaders, the agency gradually came to accept the idea, but it did so with the firm qualification that such classified areas not interfere with the agency's primary preoccupation with timber harvesting and livestock grazing. When, in 1923, Chief Forester William Greeley directed the regional foresters to map out possible wilderness areas, he required them not to include "large areas of presently or imminently commercially exploitable timber" and to ensure that the areas were suitable for controlled grazing.[91] Successive chief foresters repeated that limitation on numerous occasions.

In the beginning years of the wilderness debate this qualification on the part of the Forest Service did not seem to impede the advance of the issue in the agency, but as general ideas were applied to specific areas for management, both the advocates and the agency came to loggerheads and initial favorable agency sentiment turned into persistent opposition. Only when the Wilderness Act of 1964 turned an administrative option into a legislative requirement did wilderness become a more acceptable objective to the Forest Service. Even then, its inherent resistance to what it considered to be a "nonproductive" use of forest land continued to identify the agency as a wilderness laggard rather than a leader.[92]

The agency's approval of wilderness as a general idea appeared in the form of two sets of regulations, one being the L-20 regulation of 1929 and the other, the U-1, U-2, and U-3 regulations of 1939.[93] But it was not until these general ideas were turned into specifics by a process of reviewing lands with potential for approval as "established areas" that the implications of the general statements became manifest. The two such areas that shaped these earlier implications were the Leopold proposal for the Gila National Forest and the Carhart proposal for the Superior National Forest.[94]

The implications of the Gila proposal were relatively limited but indicated the choices that the agency might well make to undercut wilderness objectives. The area that the district forester had approved for wilderness in 1924 experienced an explosion in its deer population, and to foster easier access for hunters that might reduce it, the district forester decided in 1929 to establish a road through the area, dividing it into two parts. Leopold, who proposed the wilderness area in the first place and was no longer with the district, demurred on the grounds that this plan was contrary to the area's objective.[95] He had foreseen this possibility and in 1926 expressed the belief that "the mere embodiment of wilderness areas in Forest Service plans would not be a sufficient protection under changing personnel." As a result he favored a "bill to authorize the president to permanently suspend the operation of the land laws in areas particularly valuable for recreational purposes." The one thing needed was for the government to draw a line around each area and say, "This is wilderness and wilderness it shall remain."[96]

Action on the Superior National Forest was quite a different matter, for while the Gila decision was primarily made by the agency's District 3 forester, the Superior issue involved the agency's national policies and a variety of actions on the part of the national government. Moreover, action to protect the Boundary Waters Canoe Area of the Superior National Forest involved a national campaign promoted by Arthur Carhart, who was no longer with the Forest Service, and efforts by the American Society of Landscape Architects. The canoe area proposal was strongly supported by the relatively new Izaak Walton League, formed in 1923 and headquartered in Chicago.

The crucial issue for the Superior National Forest was a proposal by the chambers of commerce of northern Minnesota to construct roads into the forest so as to foster automobile tourism, a proposal for which the secretary of agriculture had already allocated funds provided by the Federal Aid Road Act of 1916. Thus, the campaign to "save" the Superior roadless area involved high-level decisions in Washington and a reversal of agency commitments already in place. However, action within Minnesota on behalf of the roadless proposal preempted decisions in Washington and legislation in the form of the Shipstead-Nolan Act of 1930, which bypassed the Forest Service and its evolving wilderness policy. It was no wonder that when the agency began to create specific wilderness areas under its L-20 regulation of 1929, the first to be approved was the wilderness—now called a "primitive" area—for the Superior National Forest.[97]

As the Forest Service began to approve—or, in its own jargon, to "establish"—specific areas as protected under its L-20 and U regulations, the actualities of such protected areas began to take shape. This process was interrupted by World War II but resumed in the late 1940s. As it proceeded, the objectives of both wilderness advocates and the Forest Service became clarified and the differences between them sharpened.

It became well known that the agency was serious about establishing wilderness areas only where the forest was not particularly valuable for wood production. It therefore endeavored to limit proposed wilderness areas to the higher, nonwooded elevations rather than heavily forested areas. Wilderness proponents derisively called the agency's proposed wilderness preserves the "rock and ice wilderness" and insisted that forested areas should be eligible as well.[98]

A critical point in this evolution of wilderness policy came in 1952, when Forest Service chief Richard McArdle announced that the agency would review all seventy sites that it had classified as wilderness, primitive, roadless, limited, or wild in previous years, a total of 14 million acres. Wilderness advocates soon believed that the agency's intention was to redraw the boundaries so as to exclude commercial timber, much of it old growth, and that belief seemed to be confirmed when in the next year McArdle announced that the boundaries of all seventy sites might be changed.[99]

Actions taken by the Forest Service under this review process were described by James Gilligan, one of the most knowledgeable authorities on the details of the implementation of the Wilderness Act. He wrote in 1954 that "there is a national trend of wilderness boundary modification which[,] since 1940, has eliminated over half a million acres of land from 33 different units.... These deletions have largely been offset by the addition of high, rocky zones to each area where there is little possibility of development demands or timber harvest."[100]

It appeared that the new context of wood supplies—the increasing demands of war and postwar housing and the increasing demand for supplies from the national forests in the face of declining sources on private lands—had led to the search for more sources from the national forests. This was the turning point at which wilderness advocates sought legislative guarantees that wilderness designations be placed outside the reach of the agency's administrative discretion. Their commitment to this goal was increased by the fact that during the 1930s the New Deal programs, especially the CCC, had greatly expanded the construction of roads throughout the national forests, including potential wilderness areas.

When wilderness advocates presented Congress with a measure to establish a legislatively mandated wilderness system, the Forest Service opposed it vigorously on the grounds that it interfered with its administrative authority.[101] Despite agency opposition, Congress approved the proposal in 1964 as the National Wilderness Preservation Act.

In succeeding years, the agency was faced with a more ominous turn of events as citizen groups proposed to their legislators that areas not mentioned in the 1964 act be added; these were called the de facto wilderness areas, in contrast with those designated under the law, which were the de jure wilderness areas.[102] This potential expansion of the wilderness designation would put a considerable amount of forest land into play at the behest of citizen groups rather than the agency. The de facto wilderness issue took still another turn when easterners began to propose additions to the system in almost every national forest in their region and brought proposals to their legislators, who in turn put them into the legislative hopper. The agency sought to curb the growth of these "de facto" proposals by its own survey of potential wilderness areas throughout the entire national forest system, thereby identifying areas that it thought were "eligible" for designation. This action only intensified the surge of citizen proposals and the strength of agency resistance.[103]

In response to the eastern proposals, almost all of which were on areas long since cut over and regrown, the agency developed the argument that only "virgin" wilderness was eligible under the wilderness preservation act and that there were only a few such areas in the East; thus, most of the proposed areas were ineligible. Down to the present day, citizen groups have continued to propose additions, frequently via the planning process established in the National Forest Management Act of 1976 but also in direct negotiations with legislators. However, continuing in its reactive rather than innovative mode, the Forest Service usually responded by proposing an area of far less acreage, thereby continuing the bargaining process between citizen groups and the agency.[104]

Although the Forest Service resisted the expansion of designated wilderness areas in the national forests, it gradually became reconciled to the permanence of wilderness as a legitimate land use, almost equal in importance to the uses outlined in the Multiple-Use Act of 1960. One slight but important shift in its program reflected this reconciliation. It had long argued against parks and wilderness on the ground that they called for a "no management" regimen in contrast with areas emphasizing wood production that did call for formal management.

But having responsibility for wilderness areas under the 1964 wilderness act, the agency proceeded to manage them, providing for regional supervisors and rangers to patrol wilderness areas exclusively. Under a protective rationale called the "limits of acceptable change," the agency also developed standards of use for campgrounds so as to prevent damage to soil and vegetation. In 1980, two researchers at the North Central Experiment Station in St. Paul, Minnesota, produced a manual titled *Wilderness Management*.[105]

Wilderness became even more firmly entrenched in the agency's management program when it participated with the other three main federal land agencies—the National Park Service, the Bureau of Land Management, and the U.S. Fish and Wildlife Service—in creating two centers located at Missoula, Montana: the Aldo Leopold Wilderness Center, which stressed wilderness education, and the Arthur Carhart Wilderness Center, which stressed training for wilderness management.

Despite the fact that aesthetic objectives in forest management had never received statutory authorization as a use—in neither the Forest Management Act of 1897 nor the Multiple-Use Act of 1960—the agency had provided for some recognition of this objective, either at its own initiative or from demands from the public for such things as outdoor recreation areas.

One arena in which aesthetic objectives were sharply focused was in the series of agency programs devoted to the management of the Non-Industrial Private Forests (NIPF) under the State and Private Forest Division, involving for the most part grant-in-aid programs to the states. The agency as well as the wood products industry had long been concerned about the lack of interest in wood production on the part of people who were seeking to build vacation or even permanent homes on the small tracts they purchased.

The agency and other groups commissioned a series of studies to determine the reasons for this lack of interest, and the results supported the conclusion that aesthetic forest values loomed high in land-owning objectives. By 2002, the agency had undertaken the first of what was to be a series of periodic surveys of the values that small-tract owners sought in their forest lands. In reporting on the results of the first survey, the authors posed the question, "Will aesthetic enjoyment as a reason for owning forestland continue to increase and will ownership for timber production continue to decrease?" These studies made clear that among this large group of woodland owners, aesthetic values persisted despite the desires of wood-production advocates in both private and public forest affairs to dampen them.[106]

The Evolving World of Multiple Use

During the last third of the twentieth century, management of the national forests became increasingly complex. Much of this complexity arose from the agency's new role as arbiter in the balance of uses established by the Multiple-Use Act of 1960, a task further complicated by the agency's declared conviction that, with its own expertise, it could satisfy all users.[107]

The Multiple-Use Act was not just a law that authorized the various legally acceptable uses for which the agency could manage the forests. On an even more basic level, it reflected the trend of vast numbers of individuals personally experiencing the forests; their visits shaped their views as to what the forests meant and what role they should play in their lives. In fact, this personal experience, it could be argued, played a far more influential role in the affairs of the Forest Service than did innovations emanating from the nation's capital.[108]

The new forest users tended to look upon the forests not as a resource to be extracted but as an environment that would enhance their work, home life, and play. Many, of course, experienced that environment on trips by foot or pack train into the forests, and many sought out vacation and permanent home sites that provided an environment far more enjoyable than what they experienced in the towns and cities from which they came. Viewing the forest as environment necessarily put focus on its aesthetic qualities, which the agency had never received statutory authorization to address but which it had in fact addressed on a selective basis. Now, by the very fact of public use, aesthetic objectives became a more integral part of agency affairs.

During the debate over the Multiple-Use Act of 1960, the Forest Service seemed to be interested less in the use categories that the law identified and more in retaining its freedom to decide among the various uses. This it did through the planning process, which had been gradually evolving over the years but which the National Forest Management Act of 1976 mandated for all national forests. Planning for each national forest would have to balance forest uses.[109]

While planning was officially a matter of balancing *uses,* it became in practice a process of balancing *users,* and Forest Service planning soon became marked by a clash of users, which required that agency personnel have not just technical skills but skills in negotiation, mediation, and compromise. Technical issues of forest management often faded into the background while the task of mediating conflicts among users came to the fore. Although the agency came up with the slogan "Land of Many Uses," perhaps a more accurate slogan would have been "Land of Many Users."

These circumstances within which the Forest Service had to work beginning in the 1960s became increasingly intricate because of the evolving complex of uses and users. Each one of the use categories that had arisen during the twentieth century and found its way into the 1960 law evolved into even more complex groups of uses and users, each of which had its own claim on the forests and generated another group containing an ever-larger number of users. Agency affairs after 1960 cannot be understood without a firm grasp of the ways in which user activities and user organizations surrounded the Forest Service with a maze of institutions with which it now had to work.

Recreation, for example, which in the years prior to 1960 had consisted primarily of campers and hikers, now entailed motorized vehicles, ranging from two-wheeled off-road vehicles to three- and four-wheeled all-terrain vehicles (ATVs).[110] Motorized recreation vehicles presented the agency with the same problem campers had in the 1920s. They wanted to go anywhere in the forest, but while the Forest Service had at that time brought campers under control by requiring them to camp in designated campgrounds, it was far less successful in confining off-roaders to designated trails. However, fixed recreational centers, such as ski facilities on forest land, became increasingly attractive to the agency because they returned considerable income. As the last third of the century proceeded, recreation greatly outpaced timber as the agency's main source of income.

Wilderness activities, which the Forest Service had always looked upon as a part of its recreation program, also went through extensive changes, from a first stage involving areas identified in the Wilderness Preservation

Act of 1964 to stages involving areas that were later designated as wilderness.[111] The act declared particular areas as wilderness but then added "study areas" that were to be considered by the Forest Service, which would then make a proposal to Congress for its final decision on whether to preserve the area. In the initial proposed wilderness legislation, the agency had been given the task of making the final decision as to what areas might be designated, but due to the insistence of Wayne Aspinall, chair of the House Committee on Interior and Insular Affairs, this provision was changed so that Congress made the final decision. This small shift in ultimate authority led to a new context for initiating wilderness proposals, which encouraged citizens to urge their members of Congress to introduce legislation in the order of customary constituent requests.

The initial wilderness area proposal, the Lincoln-Scapegoat area in Montana, was followed by a series of such "de facto" measures, in contrast with the "de jure" proposals identified in the law.[112] The wilderness designation process inspired citizens living near almost every national forest to draw up new proposals, which plunged the agency into a mass of wilderness-related work from which it could not escape.

When Congress passed the Multiple-Use Act of 1960, wildlife was thought of primarily as game animals. But the steady shift in perspective to wildlife as encompassing all wild animals eventually included even rare and endangered plants. This brought into the orbit of national forest management a wide range of species.[113] This broad view of wildlife challenged the Forest Service more than had the issues of recreation or wilderness, for while those activities could be segregated from the "general forest" of wood production and called "special uses," wildlife was, by the very fact of its habitat, an integral and inseparable part of the entire forest. The potential for integration of wildlife and wood production was made even more evident when the use of radio telemetry—the "radio collar"—in wildlife research tended to identify habitat used by many birds and mammals as far more geographically extensive and complex than had previously been thought.[114]

Watersheds also took on a new and more elaborate meaning. On its grazing lands the agency had long been admonished to be concerned not just with grazing activities in headwaters but also on stream riparian areas. Cattle that congregated at water holes or along streams to drink, or were brought together overnight, destroyed vegetation and made stream banks subject to erosion. While the agency had earlier concentrated on headwater erosion, it now was called upon to deal with downstream stream-bank erosion.

At the same time, an expanded interest in the decline of wild salmon migration, attributed largely to the downstream effects of upstream activities such as wood harvesting, stressed a new interest in the relationship between upstream and downstream river segments.[115] This connection got a new advocate as the U.S. Environmental Protection Agency (EPA) brought its concern for water quality to bear on forest management.

As the Forest Service took up the task of balancing uses and users, it was faced with an increasingly complex set of circumstances evolving outside the agency but having a direct impact on it. This new set of circumstances, some of which it had long ignored and many of which were new, all taxed the professional perspective and skills of the agency. These issues came into focus as the agency embarked on the task of preparing plans for each one of the national forests. Agency personnel did so with the fervent hope that they could satisfy all by increasing "outputs," as they were called, that were within the agency's technical capabilities. It expressed an unbounded optimism about its ability to resolve the human competition for its resources simply by producing more for all.[116]

At the same time, however, agency leaders generally spoke as if, among the various forest uses, wood production was still the dominant one.[117] Assistant Chief Edward Cliff, for example, expressed this view in a speech before the American Forestry Congress in 1951: "Timber production is given priority over other uses on the most important areas of commercial forest land, with recreation, livestock grazing and wildlife being integrated as fully as possible without undue interference with the dominant use."[118] One way of expressing this was to argue that although all the legally identified uses were to be "taken into account," they could best be achieved through the practice of sustained-yield wood production.

Planning as a process of adjusting uses and users evolved steadily after 1960 so that, when the 1976 act mandated that all national forests have a formal management plan, it became the central focus on the first round of fifteen-year plans to meet the requirements of that act. The agency sought to make the adjustments by balancing calculated numerical benefits through a computer program known as FORPLAN. Long-term benefits for each use were adjusted to present terms through a "net present benefit" formula. All of this simply translated the varying objectives of users into numbers that then became the focal point for contention and debate.

This new planning process also led to an important role in the planning process for a young forest economist in Eugene, Oregon, named Randy O'Toole. O'Toole was adept at identifying the meaning of the numbers in FORPLAN and provided advice to critics of the agency's own numbers. O'Toole became so adept at after-the-fact exposure of the agency's decisions that some supervisors, in order to make their numbers less vulnerable, solicited his opinion before the plans were completed.[119]

Users focusing on the forest environment complained that the numbers applied to wood production were too great and those applied to recreation, wildlife, wilderness, and watershed protection too small. The entire process, however, was overshadowed by the policies of John Crowell, the assistant secretary of agriculture in charge of the Forest Service and an appointee of the Reagan administration. Crowell, a former general counsel for the Louisiana-Pacific Corporation, the largest purchaser of timber from the national forests, believed that the agency could greatly increase the cut from the forests without adverse consequences for other resources. Crowell ordered forest supervisors to increase the allowable cut in their plans. But in doing so, he aroused the opposition of many supervisors in the West. That opposition was led by supervisors in the northern region, who protested the adverse impact on wildlife and watersheds that would result. A number of leaders of western state fish and wildlife agencies also joined in the opposition to Crowell's order.[120]

As each national forest prepared its plan, it also prepared an analysis of the plan's environmental consequences, as required by the National Environmental Policy Act of 1969. The planning process, therefore, involved two closely related documents—one allocating forest lands to different uses and users and the other investigating the effects of that use on forest resources. The first required a skill in forging compromise among users and the second required scientific inquiry into resource conditions with a view to developing some sort of conclusions about the "state of the resources" themselves.

This second task was far more than a legal requirement but reflected a circumstance that had been a direct result of the agency's tendency during and after World War II to increase resource production, that is, to foster "maximum yields" in the case of timber and "full utilization" in the case of grazing. Amid pressures to increase production, the agency frequently found ways to argue that its impact on land, water, and wildlife resources would be minimal and acceptable. The impact analysis requirements, therefore, gave sharp emphasis to these agency tendencies to compromise resource conditions in the face of user demands. This, in turn, fostered a slowly accumulating backlash of interest in the quality of forest resources often expressed under the general heading of "forest health."[121]

Budget Controversies

In the years after World War I, the Forest Service continued to struggle with the legacy of Gifford Pinchot's promise that he could "make the forests pay." In this struggle, the agency took up several strategies to defend its record in appearances before the budget committee in Congress, which, along with the timber industry, sought to improve the agency's financial record. The agency's defense strategies over the years took many twists and turns.[122]

To recap, after World War I the Forest Service justified its record before Congress not by reporting the balance between income and expenditures in normal accounting fashion but by emphasizing the increasing revenue it enjoyed, with the implication that this record overshadowed considerations of cost. It forecast a rosy future in which its revenue would ultimately cover all of the agency's costs. At the same time, it continued to divert attention from its overall financial condition with the argument that while some forests would never pay their own way, a number of them would.[123]

Increasingly, however, the Forest Service, recognizing its unfavorable financial balance from wood production alone, began to justify its program in terms of unquantifiable nonmonetary benefits. William B. Greeley, the chief forester who succeeded Henry Graves in 1920, emphasized this point in his first annual report. He called these "public benefits," which, he said, would rise in the same degree as the monetary returns from logging and grazing. These benefits, he argued, could not be quantified, and if money income alone were the main consideration, then the agency's objectives could not be realized.

In 1927, a new accounting system was instituted that would separate income-producing activities from those that were not. In 1948, the agency's report warned that growth in revenue might not exceed costs and argued that the objective of national forest administration was service and public benefits, not profits. Despite the fact that the agency resisted many of the new nonprofit activities in which it was involved, they were beginning to play a dominant role in the agency's justification of its finances. It had come a long way from Pinchot's boast in 1905 that he could "make the forests pay."

For some costs attributed to wood production, a scheme of "trust funds" was worked out in which those costs would be covered through specialized accounts with funds derived from the sale of wood and thus separated from the agency's main budget. Much like the five-year autonomy from congressional scrutiny that Pinchot had worked out in the Transfer Act of 1905 but that was rescinded three years later, these trust funds were maintained by the agency "off budget," that is, without annual budget review by Congress.[124]

In 1913, the Roads and Trails Fund was created to receive 10 percent of all income from timber sales in order to construct roads and trails in national forests. In 1916, the Brush Disposal Fund was created to pay for burning the brush and slash from harvest operations; proceeds from the sale of timber went into that fund. In 1930, Congress created, as part of the Knutson-Vandenburg Act, the Replanting Fund, in order to fund replanting on logged-over national forest land. Finally, in 1976, Congress created the Salvage Sale Fund, with a one-time appropriation of $6 million for the purpose of removing dead and damaged trees from the national forests, with the added provision that forest administrators could deposit all revenues from salvage sales back into the fund rather than returning them to the U.S. Treasury.

The crucial feature of these trust funds was that the agency did not have to account to Congress for their income or expenditures but supervised them solely on its own. Over the years they came under considerable fire from citizen taxpayer organizations, who urged that they should be subject to annual budget scrutiny by Congress. This concern arose when it was charged, and substantiated by the General Accounting Office, that the agency was using the funds for administrative expenditures such as salaries and equipment.[125]

Of longer-term significance was the agency's accounting of nonmonetary activities such as wildlife and wilderness management, watershed protection, and recreation, whose financial implications became an important feature of the debate over the Multiple-Use Act of 1960. The more the agency used the benefits of these activities to justify its financial performance, the more the timber industry sought to find some way to reduce their role in the program of the Forest Service. In the proceedings leading to the 1960 act, the industry sought to delete from the list of "uses" those added after the original two (watersheds and grazing) identified in the Forest Management Act of 1897. At the same time, it argued that the national forests were vastly underutilized, could produce far more timber and thus revenue, and, if confined to that objective, could pay for themselves.

The National Lumber Manufacturers Association submitted a proposal to make wood production the agency's primary objective and make the agency self-supporting as much as possible. In response, the Forest Service proposed a provision that it was not necessary to secure "the combination of uses that will give the greatest dollar return or greatest unit output." This wording became part of the Multiple-Use Act of 1960.[126]

Some commentators argued that this vague and general statement gave the agency carte blanche to avoid normal accounting. But it was far more important as a step in applying cost accounting to the entire range of forest uses. In the legislative proceedings, this process of altering the agency's accounting practices took a rather ominous turn when the committee asked the

Forest Service to define the various identified "uses." The agency declined to do so. In retrospect, even this exchange moved the agency a bit toward not only defining costs but also putting numbers on them, a task that it had long avoided. In planning following the National Forest Management Act of 1976, the agency developed the format of balancing a range of forest uses in terms of their quantified "benefits" and ushered in a new agency accounting era in which the arguments over the balance of uses were cast in terms of the selection of the appropriate numbers used to make the calculations.

Thus, despite the agency's reluctance to advance the role of wildlife, recreation, watershed protection, and wilderness in its program, it increasingly relied on their benefits to justify its finances and now was under increasing pressure to quantify those benefits in the tangible numbers of an accounting system responsible to Congress and the public.

Reaching Out for Public Support

In the years after World War I, the Forest Service developed a favorable public reputation as an agency of ability and competence, but at the same time it suffered a political weakness resulting from the agency's inheritance of a "management from the top" philosophy. This was an outgrowth of the decision-making strategy of the Theodore Roosevelt administration combined with Gifford Pinchot's alienation of much of the public interested in the forest reserves when he excluded their objectives from his management program. After World War I, the agency could not, for example, call upon the sportsmen's organizations or the women's clubs who constituted the agency's largest potential base of support because of the direction Pinchot had set. A number of developments at that time reflected the agency's limited public support amid recurring political crises.

One group with which the Forest Service sought to establish a working relationship was the Sierra Club. The result of the agency's efforts was an off-and-on "love-and-hate" relationship, which continued until a sharp divorce took place shortly after World War II.[127] The Forest Service sought such political support in the face of a variety of challenges: fighting the ongoing effort of the National Park Service to secure agency lands for new parks; minimizing mining, an activity occurring in forest lands but over which the agency had only limited influence; and reacting to the pressures in Congress either to privatize the forests or transfer them to the states, a proposal championed by President Herbert Hoover. The agency considered the Sierra Club to be a useful potential ally because of its strong support for the reserves as public assets at a time when often it was the very principle of public ownership of the forests that was at stake. While the club differed with the agency on many specific issues, it tended to take a pragmatic role, working with the Forest Service on some issues and opposing it on others.

The relatively permanent divorce between the club and the agency took place as the Forest Service in the early postwar years took steps to revise its various protected area classifications. This move made clear that the wilderness, wild, and primitive areas were in no sense permanent and that under the pressures of increased wood production the agency would be more than prone to open some to wood harvest. Under the vigorous action of its newly established executive, David Brower, the club attacked the agency's new direction and pressed Congress to place the designation of wilderness areas beyond its control. Agency initiative and club reaction set the beginning stages of the permanent divorce and the end of the agency's effort to rely on the club for significant political support.

The most successful Forest Service attempt to reach out to the public and gain its support came from quite a different quarter—its drive to enlist the public in preventing forest fires. This effort included the development of the "Smokey Bear" media campaign, a public relations strategy that had remarkably good results. This was a fortuitous event in a fire prevention campaign that brought federal, state, and private interests in wood production together for several years and undertook a major public relations push to offer dramatic lessons on how people, through such careless acts as throwing away lighted cigarettes or failing to put out campfires, were major causes of forest fires. The campaign centered on the image of Smokey Bear admonishing all who viewed and listened that "only you can prevent forest fires."

Through the use of print media and radio, the message was spread widely, promoted by the Advertising Council and a variety of civic organizations. A historical record of such appeals—remarkable in its variety and range—was created by the Forest Service centennial history project in two discs of "bonus material" accompanying the main DVD.[128] Smokey Bear was, so the history has it, the second most recognized symbol in the public media. He created a vast amount of fan mail, so much that the U.S. Post Office created a separate ZIP code for him in Washington.

There was a significant twist for the Forest Service in the Smokey Bear experience. In dramatizing the impact of forest fires, Smokey spoke on behalf not only of the trees but also of his "friends" in the forest, depicted as a wide range of lesser species. Fire threatened all of them; the use of cartoon characters shaped the message to appeal especially to children. The "only you can prevent forest fires" campaign not only warned against fire but also, in doing so, gave considerable impetus to the image of the forest as a wide-ranging system of biological life, something that the emphasis on wood production as the "dominant use" had tended to obscure. But amid the growing interest in the forest as a diverse biological system that emerged in the latter part of the century, the anti-fire Smokey Bear ads made a significant contribution. The agency would soon transfer the drama of this threat—fire—to other threats that were given more attention in later decades.

It is worthwhile to contrast the Forest Service's varied initiatives to develop a reliable clientele with those of the National Park Service. The approach of the two agencies differed sharply. The Park Service worked aggressively to develop a loyal clientele with its interpretive programs (at campfires and for field trips), its educational displays at visitor centers, and its network of natural history associations, all constituting a major outreach program focused on individual people becoming connected in one way or another with individual parks. All of these efforts helped to generate a broad public interest in the parks.

The Forest Service, on the other hand, took up such efforts far more slowly and in limited fashion. Faced with a public that was skeptical about wood production as the primary meaning and use of the forest, the agency was prone to admonish rather than encourage, to emphasize that the public did not recognize the difference between parks and forests or did not understand that, following an unsightly harvest project, a forest of aesthetic beauty would emerge.

Ever since the automobile became widely used, a vast number of Americans had come to know and enjoy the forests in recreation and nature observation, and they constituted a clientele of enormous potential value to the agency, one on which it might rely for support. But while the Forest Service was quite prone to engage in vigorous competition with the Park Service by building recreational facilities, it did not look upon its visiting public as an opportunity to build the clientele that it had long sorely needed.[129]

The Legacy of an Expanded Clientele

The Multiple-Use and Sustained-Yield Act of 1960 provided statutory legitimacy to those forest uses and users that Gifford Pinchot early in the twentieth century had excluded from the Forest Service program. They sought to reverse that policy after World War I, and had thus become participants in forest affairs. Wildlife, watershed, and aesthetic objectives became an integral part of the agency's responsibilities. Despite this statutory acceptance, the agency did not welcome the influence of their advocates in its affairs.

During the debate over the 1960 act, the Forest Service seemed to be interested less in the problem of spelling out the range of uses and users and more in its ability to retain authority over establishing the balance among them. Agency leaders spoke frequently in terms that made clear their view that, among the various uses, wood production remained the dominant one and the others, though legitimate, were subordinate to it. Another way of expressing this ranking of uses was the argument that, while all were recognized, it was only through a successful wood production program, from planting to harvest, that the other uses could be achieved effectively.

The future was to be a rocky road for the Forest Service. It was now responsible for accommodating a group of uses and users that it had admitted to its management objectives only reluctantly. They made their own demands as to how the forests were to be used, demands that the agency was ill prepared to face because of the limited training and perspectives of its staff. These demands, moreover, were not restricted to the meaning of such categories as recreation, wildlife, watershed protection, wilderness, grazing, and wood production at the time of the 1960 act but were greatly augmented due to the fact that the categories of "use" and "user" expanded into a wide range of different activities, each one of which had its own particular twist.

In coping with the new world of multiple-use objectives, the Forest Service was faced with contrasting directions in forest management. One involved pressures from the wood and meat markets to increase production of both with greater "inputs" from production science and technology. Both markets prompted the agency to move away from its earlier ambitions of long-term silviculture and adjust the use of the range to its inherent limitations.

Another management direction, fostered by recreation, wildlife, and watershed objectives, gave rise to the use of the forest as an environment, an aesthetic environment on the one hand, a habitat environment for plants and animals on the other, and the watershed as a hydrological network of the many and varied components of streams. Because of its commitments to wood and forage production, the agency found it difficult to think of the forest as an environment for all of the natural world.

The years after World War I brought to the fore one of the agency's most celebrated employees, Aldo Leopold, who expressed a quite different vision. Very early in his career with the agency, before World War I, Leopold wrote in the *Pine Cone,* a newsletter of the Carson National Forest in New Mexico where he worked, his views about Forest Service objectives: "We are entrusted with the protection and development, through wise use and constructive study, of the timber, water, forage, farm, recreative, game, fish, and aesthetic resources of the areas under our jurisdiction. I will call these resources, for short, 'The Forest.'"[130]

Leopold's view was, in effect, a challenge to the agency to think beyond the limitations that Pinchot had imposed on its mission. But the agency did not take up most of the broader management program that Leopold had identified. While he sought to implement his vision by drawing up management handbooks for wildlife and for watersheds that his fellow employees could use, the agency was more than lethargic in moving those objectives into significant management programs.

Leopold's later role in the agency's self-constructed history was paradoxical. In those accounts, it placed Leopold's advocacy of wilderness high on its list of agency achievements in such a way as to obscure its continued resistance to advances in the wilderness agenda. And in that history it gave little support to his contributions to wildlife and watershed issues, which were equal if not greater contributions to forest management objectives.[131]

Not many years after the Multiple-Use Act of 1960 became law, wildlife, recreation, and aesthetic objectives became more explicit through specific statutes and as a result of extensive interest of the public in Forest Service affairs. A new environmental and ecological context emerged in the world of established uses, all to make the agency's task more complicated and to reveal just how poorly it was prepared to face this new world of forest management.[132]

Thus, the Forest Service entered the last third of the twentieth century with a more expanded legacy from its earlier years. It promised a new age for the agency. Little of this could have been predicted from the details of the Multiple-Use Act of 1960, which consolidated the uses that had emerged in the preceding forty years, all of which had been shaped by influences from outside the agency. Within the new framework of the 1960 act, those external influences, increasingly more complex, defined the choices that the agency confronted as it faced the immediate future.

4 CONFRONTING THE ECOLOGICAL FOREST, 1976-2005

DURING the last third of the twentieth century, a combination of initiatives from citizens and scientists brought a new perspective to bear on forest management. From one vantage point it brought together components of wildlife, watershed, recreation, and aesthetic objectives, most of which had been subordinated to extractive agency commitments in grazing and wood production over the years but now were brought to bear more forcefully on forest management. From another vantage point it reflected the broader public pressures associated with the evolving ecological perspective and represented by the National Environmental Policy Act (NEPA), the Endangered Species Act (ESA), and the substantive environmental and ecological issues inherent in the National Forest Management Act of 1976 (NMFA). Much of this came together during the 1970s and 1980s in a complex of closely related objectives that added up to an ecological approach to forest management.[1]

Coming to the fore were new implications for the debate over the role of the national forests in American society. Ecological objectives in national forest management reflected a social trend beginning in the early 1920s in which vast numbers of individuals personally experienced the forests and in the process shaped their own views as to what the forests meant and what role they should play in their own lives.[2] Because they used and experienced the forest environment firsthand, they sought to become a more integral part of agency affairs.

During the 1970s, environmental awareness of the forest began to be infused with an emerging ecological perspective that had both popular and scientific components. Scientific curiosity about the workings of the natural world joined with an aesthetic appreciation of its beauty to establish a popular affinity for all things natural in the forest. Appreciation of the environmental forest arose initially from its use for recreation. But in the 1970s, this experience began to be augmented by a new experience of the forest as an ecological system that was increasingly intriguing.

The forest as environment and as an ecological complex blended a human experience that combined both aesthetic appreciation and scientific curiosity. The aesthetic quality of forests had long been a factor in what Americans had thought to be important in forest management, even though the agency had not received statutory approval for such objectives and agreed with them only tentatively and selectively. In the past, the articulate advocates of aesthetic values had been few. Now they were many, a reflection of just how large was the number of people who had come to know the forests firsthand.[3]

The scientific curiosity that the forests now aroused was a newer attraction, a product of the ecological awakening that occurred after World War II and that brought significant sectors of both the public and the scientific community together to understand better the complex ecosystems that the forests constituted. While the aesthetic qualities of the forest environment were more readily observable, forest ecological circumstances were less so. They were often called a "secret forest," which made studied observation and investigation an attractive preoccupation.[4]

Beginning in the 1970s, this vision of the forest as a complex of ecological elements and relationships began to affect the agency.[5] It became an important element in the second decade of planning under the 1976 act, and it joined citizens and scientists in parallel though not always formally cooperative ventures. While in one sense it brought a new "ecological vision" to compete with the agency's long-held "silvicultural vision," in still another sense it was the first serious attempt to integrate those elements of forest management outlined in the Multiple-Use Act of 1960 with the mandates inherent in the NEPA, the ESA, and the NFMA of 1976. And it meant that the Forest Service would enter its second century with a significant addition to the legacies it had inherited from the more distant past.[6]

The increasing interest in nature was reflected in a wide variety of areas: elementary education, television programs, and nature photography and coffee table books, as well as attendance at outdoor-oriented museums and participation in field trips. These experiences brought closer attention to the species composition of forests, their plants and animals, and to more disciplined study and understanding of them as well as personal enjoyment. In the 1970s and 1980s and into the final years of the century, the national forests, along with other federal and state lands, came to be thought of as distinctive reservoirs of the natural world, which was increasingly scarce in the urbanized world of daily life.

To the professional forester and the U.S. Forest Service, this surge in interest was almost incomprehensible, being far different from their training and considerably beyond their experience in forest management. The agency had usually rejected this focus on ecology as not within the scope of its legitimate affairs. Not being understandable, ecological forestry was looked upon as a threat to agency values, its accomplishments, and its role in American society. The Forest Service had confined itself to a narrow slice of the forest biological world and now it was unprepared to consider as legitimate the ecological world evolving around it.

One of the early nationwide expressions of the environmental/ecological vision of forest management appeared in a survey of public attitudes toward the national forests sponsored by the American Forest Institute in 1977, shortly after passage of the National Forest Management Act of 1976.[7] The survey found that the public placed wildlife, watershed protection, and an appreciation of naturalness at the top of the list of preferences in the value of the forests and wood production far down the list. Similar studies were repeated in later years.[8]

A significant aspect of the public's attitude toward the natural world concerned its importance for future generations. Attitude studies revealed an interest in both the public's current ability to experience and appreciate nature and an "existence value" of the natural world. Even more important, however, the natural world was viewed as having a "bequest value," something that people wanted to pass on to their children.

Soon the Forest Service itself took up attitude or "value" surveys, with similar results. A survey conducted for the agency's southeast region (the southern Appalachian forests) in the year 2000 produced this summary of the results for the region: "[the forests] are viewed as important for their importance in passing along national forests for future generations, followed by protecting sources of clean water, providing protection for wildlife and habitat, providing places that are natural in appearance, emphasis on forest health, and protection of rare or endangered species. . . . The people [who live here] clearly put ecosystems and naturalness above utilitarian objectives in the management of these national forests."[9]

Forest science went through a similar development. Few scientists were associated with the growing perspective of the forest as an aesthetic environment. However, an ecological perspective was quite different. As ecological scientists grew in numbers and activity, in about the same fashion as did ecological perspectives among the general public, they began to voice views about the management of the national forests in a fashion somewhat similar to that of the public.

In contrast to those scientists who explored the intricacies of wood production, who were associated with the "forestry profession," and who were brought together in the Society of American Foresters, ecological scientists were associated with a diverse number of biological disciplines, maintaining connections through such organizations as the Ecological Society of America, the American Institute of Biological Sciences, and the Natural Areas Association. At the same time, many ecological scientists became involved in the national forest planning process as technical experts in informal though important association with citizen groups to express new ideas about how the national forests should be managed.[10]

These activities on the part of both citizen and scientist reflected a recognition of the importance of a vast array of forest biological life, which, while always there, had long been only barely recognized. In the East, much of the diversity of plants and animals, the birds and wildflowers, the amphibians, the fungi and mosses, had been destroyed through extensive logging, but with the second-growth forest the diversity was now becoming more visible and more easily discovered, especially through exploration of the few older forests that had survived that clear-cutting onslaught.[11]

In the West, those older forests that had survived the initial ambitions of the professional foresters to eliminate them as quickly as possible were explored with equal fervor.[12] These "secret" forests, which the silviculturalists had bypassed because of their intense concentration on a few species of trees with commercial wood fiber potential, were now discovered by society in general, professional biologists, and citizen groups. Having gained visual access to the "external" forest through outdoor recreation activities, the American people were now discovering its "inner" workings and insisting that this interior aspect was of as great if not greater public value than wood resources of the forest.[13]

At first, those who brought the ecological message to forest affairs were simply considered another interest group to be thought of in the same way as the groups that advocated the various multiple uses covered under the 1960 act. But the ecological approach to the national forests was something quite different, representing not another set of group interests but reflecting values deeply rooted in society as a whole. In some respects the ecological approach was closer to the long-held but vehemently rejected aesthetic view of the forests, since many involved in ecological affairs believed that a combination of scientific discovery and aesthetic appreciation of the various elements of the inner forest enhanced respect for the natural beauty of the environmental forest.[14]

Suffice it to say that as the end of the twentieth century approached, an overriding new task of the Forest Service was to forge a creative response to the implications of ecological forestry.[15]

The Elements of Ecological Forestry

National forest management plans in the 1980s had focused primarily on the balance of uses outlined in the Multiple-Use Act of 1960. But planning in the 1990s took quite a different turn; it was much more involved with the various elements of the now active ecological forestry issues. From the late 1980s onward, recently organized citizen groups, quite different from those representative of the 1970s and 1980s, participated in the formation of forest plans; these groups were organized at the level of each national forest and were more involved with each forest as part of their own "backyards."[16]

At the same time, scientists had not been much involved with the multiple-use issues of the earlier decades, but they now became much more interested in the ecological implications of forest management in the 1990s plans. They represented a variety of biological and geological scientists: botanists, wildlife ecologists, hydrologists, zoologists, and archeologists. They were quite different in their scientific activities from the traditional commodity-based silviculturalists, were housed in different academic departments and professions, were members of different professional societies, and published their research in quite different professional journals. Each found that the national forests provided an appropriate location for their scientific investigations and therefore took up an interest in their management.[17]

Much of this activity came together in the late 1980s around the relatively new concept of biodiversity. While biodiversity was often formulated in systematic phrases such as "the variety of life on earth," which was being threatened by ever more intensive development, its significance lay simply in the relatively new recognition of the enormous range of animal and plant species in the natural environment and the intricate relationships among them that formed an ecological system or "ecosystem." This new focus seemed to bring together a wide range of human experience, excitement in discovering this natural world more precisely, and aesthetic appreciation of its order, disorder, and workings. While this term had not been a part of public discussion of natural resource affairs prior to the 1990s, deference to biodiversity in the last decade of the century became an almost mandatory ritual.

A variety of forest management venues provided opportunities for those with ecological objectives to express their views, and the resulting issues gave rise to the specifics of ecological forestry. These included participation in drawing up forest plans, appealing them within the agency, and pursuing court cases. They were most extensively expressed in the citizen-formulated plans, alternatives to those developed by the Forest Service, that served to highlight the difference between agency and citizen. All this generated a series of issues over the years that made up an ecological forest agenda.

The building blocks of ecological forestry were the forest species that had long been obscured by the concentration of the wood production specialists on those few tree species with commercially valuable wood fiber. Ecologists estimated that forests contained some fifteen thousand species all told, while those of interest to the foresters were no more than several dozen or less than 1 percent of the total. Innovations in scientific methods facilitated their identification and study. Radio telemetry tracked the movement of large mammals such as grizzly bears and lynx and was a particular boon to the observation of the flight of birds. Advocates wondered how to persuade the Forest Service to incorporate at least some of these species into their management, their budgets, and their professional skills.

The agency was required by law to act on behalf of threatened and endangered species by the Endangered Species Act of 1973, and compliance would require species inventories. Botanists, ecologists, and zoologists were added to the staff, but they seemed less involved in comprehensive species programs than in strategies to prevent the agency from neglecting its responsibilities under the law. The agency found it difficult to go beyond defensive strategies, take a more proactive stand, and direct resources to inventory, manage, and monitor ecological resources and ecological processes forestwide.

Much more difficult for the agency was a management program for species other than those protected under the 1973 act. An initial step in their management was provided for in the National Forest Management Act of 1976, which required that the agency maintain a diversity of species. In the regulations developed in 1982 to implement that act, the agency was required to maintain the viability of a wide range of plants and animals in the forest, with the stipulation that no harvest activities take place without a prior species inventory from which the impact of harvest on species could be determined.[18]

Management to maintain species viability, or prevent population decline, required inventories as well as monitoring, and the agency was not prepared to devote to those tasks the necessary amount of financial and staff resources.

Over the years, while citizens and scientists pressed the agency to pursue an effective species viability program, it declined to do so. In sharp contrast with the National Park Service, the Forest Service showed little interest in systematic species inventories. While the Park Service was undertaking a ten-year "all taxa" inventory of the Great Smoky Mountains National Park, the most comprehensive Forest Service biodiversity inventory consisted of vascular plants on the thirty-seven-hundred-acre Massabesic Experimental Forest in York County, Maine.[19]

A focus on the wide range of forest species inevitably led to thinking about the forest as a "home" or habitat for them. This was a more difficult program for the agency than was recreation, which required only that the agency look upon the forest as a pleasant environment. Recreation objectives could be satisfied by use of a zoning procedure in which land and water were allocated for recreation in aesthetically pleasing forest areas. But species habitat could not be so easily segregated from the "general forest," which was the term the agency used for that part of the forest devoted to wood production. The forestry profession and the Forest Service did not think in terms of habitat and, in fact, argued frequently that no special biological framework was needed around which to organize data about species habitat.

The basic unit in forest management, the species "stand," could readily provide all the information about habitat that was needed.

Advocates of ecological forestry, however, took a different view of habitat. Habitat classification required quite a different set of categories than did forest area classification for wood production. There were wetland habitats and riparian habitats, ridge habitats and cove habitats, early successional habitats and old forest habitats; there were mixed-age forest habitats and even-aged forest habitats; there were large, deep forest habitats and fragmented forest habitats. Once one focused on species, it was quite logical to carve up the forest into units directly related to their life rather than just the life of the harvestable trees.

Radio telemetry greatly expanded the territorial scope of habitat by identifying not only areas where animals produced young but also areas where animals foraged for food and adult birds taught their young to fly.[20] One could also distinguish the habitats required by wide-ranging terrestrial animals and free-roaming and migratory birds from those of animals that roamed only over short distances or plants that remain stationary and are pollinated by insects and birds. All of this was a world of common understanding and knowledge to the ecologist but beyond the experience of the traditional forester, and it constituted an "ecological imagination" that was far different from the "silvicultural imagination" of the professional forester.

Watershed Management

Ecological forestry advocates brought into their policy proposals a serious effort at implementing an agency watershed program. For almost a century, the watershed mandate in the Forest Management Act of 1897 had been subordinated to other objectives by the Forest Service to the point of almost total obscurity. The problem was that its more vigorous grazing and wood production programs often came into conflict with watershed objectives, and, in making choices, the agency opted for resource production rather than watershed protection. While the agency continued to include its watershed responsibilities in its public statements about forest management objectives, it continued to give them no distinctive administrative identity and to provide no specific management strategies to achieve the goal of watershed protection.[21]

In the last third of the twentieth century, the role of the national forests in watershed protection took a more positive turn. This change arose in part from the nation's water quality program administered by the Environmental Projection Agency, the standards for water quality which it developed and sought to apply, and especially its slowly and gradually developing interest in "non-point" sources of water pollution, or that which does not come out of discrete outfalls. This led to the EPA's frequent evaluation of national forest plans and projects and its focus on the agency's shortcomings in watershed management.[22] It also stemmed from the workings of the Endangered Species Act of 1973, which applied to aquatic as well as terrestrial species and led to an emphasis on the negative effect of land use practices, including forest management, on the fate of fish. That relationship was given special visibility in the Pacific Northwest in the case of wild salmon, which depend on high-quality waters for spawning. All of this became focused on the issue of the degree to which water pollution threatened the survival of major salmon species.

Another circumstance that gave rise to more positive watershed action was the discovery of the process of atmospheric acidification, known as "acid rain." This discovery arose from advancing knowledge of the fate of sulfur dioxide emissions from electric power plants. These emissions were carried up into the wind currents, transformed into sulfuric acid, and deposited in rain and snowfall onto downwind waters and vegetation.[23]

The ensuing debate over the extent of the effect of acid rain on forests generated an intense interest in the subject on the part of citizen groups and scientists. But the Forest Service and professional foresters in general were quite skeptical and tended to argue that damage to forests was minimal. While considerable interest in acid rain arose from outside the agency, the agency itself, save a few exceptions in some regional offices, was looked upon as lagging rather than leading in the issue. Watershed chemistry became even more complex when some scientists conducting "whole watershed studies" began to focus on the loss of nitrogen in harvested forest soils and its buildup in streams, an aspect of stream water quality that in past years had received little attention.[24]

Ecological forestry brought several aspects of watershed management into focus. There was a particular emphasis on ways of thinking about watersheds as a whole in assessing the overall condition of the network of water courses from the initial headwater streams, which ran only seasonally or intermittently to the lower perennial streams that ran continuously. The focus on whole watersheds emphasized several issues: the aggregate effect of many "incidents" on the same watershed, the cumulative effects of "incidents" over time, and the need for cooperative action when a single watershed was divided among several public and private landowners.[25] Basic to all of this was the lack of ideas as to how to think about and manage whole watersheds, the lack of data to assess their condition, and the lack of professional skills to guide an effective policy. Forest hydrology management personnel came only slowly to the agency, and there appeared to be no special division through which watershed objectives were organized. At the end of the twentieth century, watershed affairs seemed to be confined to the lowly Watershed, Fish, and Wildlife (WFW) office in the Washington headquarters, which established communication networks with a few specialists in the regions but did not have a management program. Because of the long-term neglect of its watershed responsibilities, the agency was ill prepared to respond effectively to new interest in the condition of forested watersheds.

Watershed issues such as these began to appear in the response of citizens and scientists to specific timber harvest projects, in their critiques of forest plans, and in the alternative plans they prepared for individual forests. Watersheds also became issues at the state government level, especially in the administration of forest regulations in California, Oregon, and Washington. Some forest harvest projects were challenged on the grounds that they increased soil erosion; in other cases the Forest Service was challenged when it argued that while one project on one watershed might increase erosion, that problem would be offset by watershed improvements on another. Often the agency made statements about the effects of harvests on watersheds that turned out to be assumptions when data to back them up were lacking.

In California the ecological forest watchdogs advocated that harvest proposals include water quality monitoring before and after harvests and that maps of earlier projects be provided for each proposed harvest to help assess the cumulative effects on watersheds over time.

Advocates of a more effective watershed protection program were especially pleased with the program for the Pacific Northwest forests developed by the Clinton administration in the 1990s. It was driven by endangered species issues involving aquatic species, especially the wild salmon, whose decline, it was argued, was due at least in part to the decline in water quality arising from logging on the headwaters of Pacific Coast streams.[26] The Forest Service traditionally met such criticism by suggesting that the detrimental effect of harvest on watersheds was minimal, whether it was affecting drinking water supplies or aquatic habitat. Thus, to cities that drew their drinking water from the national forests, the agency continued to argue that its harvest policies would not harm the quality of their water supply. In new rules formulated by the agency during the George W. Bush administration, the agency continued to minimize such effects.[27]

Rather than develop its own standards for watershed management based on the quality of whole watersheds, the Forest Service tended to rely upon "best management practices" (BMPs), water quality standards which emerged in the states as a result of cooperative action among timber companies and state forest agencies. These were voluntary arrangements emphasizing measures to reduce sedimentation created by timber harvesting. In a few cases, these measures took the form of state statutes, which were poorly enforced. But when the adequacy of the Forest Service program for watershed management was called into question by the federal Environmental Protection Agency, the Forest Service took refuge in the BMPs, calling them sufficient even though for the most part they were either unenforceable voluntary arrangements or only weakly enforced.[28] And in a number of southeastern states, it was often noted, forests were exempted from the jurisdiction of the state water quality programs.

The forest plan for the Northwest contained two features that advanced an effective watershed program and served as models for other areas. One feature was that the endangered species issues cut across the property lines, whether the landowners were different federal agencies or federal-state and public-private administrations. Thus, the interagency program for watershed management in the Northwest plan was able, temporarily at least, to shift the watershed focus to the watershed itself. The plan's other feature was to foster the creation of a watershed-wide monitoring system, integrating hydrological information about all elements of the watershed, upstream as well as downstream, into one overall watershed assessment. An inter-agency committee established by the Northwest program developed the first watershed-wide strategy for inventory and monitoring, the absence of which had long marked the agency's neglect of its watershed responsibilities.

Both of these major advances incorporated in the Northwest forest plan, however, were short-lived, as the subsequent Bush administration reduced support for them and they languished.

Habitat Disturbance and Restoration

The cluster of species/habitat and watershed objectives that were an integral part of ecological forestry initiatives were augmented by a series of "disturbance" and "restoration" proposals that involved management actions to reduce harm to forest ecosystems and to restore those already damaged. These actions tended to involve parallel and often cooperative action among citizens and scientists as significant as those involving species/habitat and watersheds. The agency's neglect of disturbance issues was made worse by its preoccupation with maximum production of timber and forage, which had been stimulated by World War II resource demands that continued unabated after the war and tended to generate minimal attention to resource conditions.

At the same time, the attention given by the Forest Service to restoration was shaped heavily by the agency's longtime preoccupation with wood production. Restoration from past resource damage consisted largely of re-creating the "silvicultural forest" rather than the "ecological forest."

In the agency's early years, when recreation in the national forests was closely linked to public appreciation for their natural beauty and the desire to enjoy a short period of quiet and escape from urban development, clutter, and noise, the public also got a taste of forest "disturbance." The main threats of disturbance to the aesthetic experience were roads and the sounds of motorized travel. But as the interest in the forests evolved from a recreational experience to a focus on the forest as an ecological system, the idea of forest disturbance became extended to the condition of the forest ecosystem, declines in the population of species, and the disruption of ecological relationships among them.

Along with this change in the substance of forest disturbance came a vast increase in the intrusions from the world beyond the forest, from developments and roads to a wide range of off-road motorized activities, to visual impairment from sources such as electric power plants outside the forests, acid rain deposition, and mining within the forests, all of which generated a series of threats to ecological conditions that otherwise were subject only to natural processes.[29]

Thus, ecological forestry tended to expand the meaning of "disturbances" from disturbances in the recreational experience to disturbance in ecological forest conditions. And this, in turn, tended to give rise to a working relationship between the citizen interest in ecological forest conditions as a circumstance of popular scientific interest and natural beauty and the ecological scientist, having similar interests but more focused on gaining scientific understanding of the forest as an ecological entity with its wide range of species and its intricate interconnections and patterns of change.

In earlier years, when the recreational experience seemed to dominate the citizen interest in the national forests, ecological disturbances generated little concern on the part of the scientific community. In the age of forest ecology, however, a far wider range of scientific investigations and thus of scientists became involved in national forest affairs and made their own contributions along with citizen groups in debates over appropriate forest objectives. They made joint appeals to investigate, inventory, and monitor the wide range of species in the forest and to identify with equal care the impact on those species from timber harvest, recreation, grazing, road building, and acid rain.

The range of forest applications for the "ecological imagination" fostered by ecological scientists was often broader than that of citizen groups and tended to stretch their perspective. One example of these broader scientific concepts was "forest fragmentation," the process by which the larger, intact forest was fractured over the course of forest history by roads, divisions in ownership, power lines, and informal access routes, thus extensively disrupting forest ecological relationships.

Fragmentation was especially harmful to those species that required large, undisturbed areas; it reduced their populations while fostering those species that could thrive on smaller habitats characteristic of human habitation. This broader perspective on the scope of human disturbance to natural systems was beyond the immediate human experience that had energized much citizen activity, but included it. Especially significant was the way in which ecological scientists brought to the table the condition of the "lesser" species, such as invertebrates, mosses, lichens, and amphibians, which were far less visible in the media-influenced experience of the general public than were the large mammals.

Forest restoration management was closely related to the debate over forest disturbances. It also highlighted differences of opinion on restoration objectives. The benchmarks to measure the success of ecologically based restoration differed markedly from the benchmarks of tree regeneration that were central to commodity forest production. In traditional forest culture, the main problems were forest diseases, insects, and fire, all threats to the growth of forest trees for commercial use. Now, however, ecological forestry advocates brought to the table a much wider range of ecological problems (many resulting from more intensive forest uses) that had arisen over the years and that required attention. Soil erosion, the condition of watersheds, and species declines all involved problems that, to the commodity foresters, were problems of minimal importance. Therefore, they were not high on the list of objectives when it was time to assign funds and professional attention.

Because ecological scientists viewed the forest as an ecological system rather than as primarily a source of marketable wood products, they brought to forest issues a wider range of objectives in forest restoration, and citizen forest reform groups joined with them. They stressed the need to reduce the forest road system, not simply to halt the construction of new roads but to reduce existing road mileage. They emphasized the need to restore the mixed-age forest characteristic of earlier years, prior to the even-age forest management promoted by industrial forestry, which, according to both scientist and citizen, reduced the diversity of habitat and species. They supported measures to halt the decline of species, to identify the threats to species populations from forest disturbances such as roads and wood harvest, and to restore those that had declined. They urged the agency to think in terms of enhancing habitat for forest species and of protecting watersheds. They urged the Forest Service to achieve those purposes by reducing the timber harvest to levels that would foster forest ecological restoration, thus subordinating wood harvest to ecological objectives rather than vice versa.

Debate over the problem of forest restoration was sharpened in the early twenty-first century by the results of fire and the question of forest practices following a fire. Should the burned timber be salvaged for its wood fiber or should it be left to provide the beginnings of a restored ecosystem? The issue was highlighted by two fires, one in the Bitterroot National Forest in Montana and a much larger one in the Siskiyou National Forest in southwestern Oregon and northwestern California.

The second fire, known as the Biscuit fire, gave rise to a proposal for a very large salvage log cut and a counterproposal by ecologists allied with citizen groups and a biologist for the Forest Service who argued that a salvage operation would damage the burned ecosystem and retard forest restoration.[30] The issue involved two quite different ideas as to what a forest was all about: was it a source of marketable wood fiber or was it an ecosystem of many species and ecological conditions that required more comprehensive understanding and attention? The salvage proposal prevailed, but with far less attention to the retention of charred trees to protect streams and minimize road building.

Disturbance and restoration issues came to be thought of in the media as issues of "forest health" with an overriding emphasis on forest fires and the consequences of fire. *Forest health* was a term widely used to identify the desired condition to which forest management should aim in identifying recovery from damage. But while the ecological foresters brought to the issues of forest health the objective of establishing a particular forest condition, managers of commodity forests viewed forest health according to the condition of trees commercially valuable for their wood. And in the public debate this wide range of ecological forest conditions was obscured by the overwhelming importance of fire.[31]

The controversy underscored two aspects of the role of fire in forest health: the role of an overgrown forest of small but stunted trees in fueling more severe forest fires and the longtime practice of suppressing all fires, thus fostering the growth that made forest fires worse. As a result, the Forest Service began to change its view that all forest fires were undesirable and to develop "controlled burns" to reduce the growth that fueled forest fires. Ecological forest advocates agreed with this practice; at the same time, they sought with far less success to provide an ecological focus to forest health that would go far beyond fire and extend to desirable forest-wide ecological conditions.[32]

Implementing Ecological Forest Objectives

The Forest Service had mixed reactions to the concept of ecological forestry. While the agency occasionally recognized the changing views of the public about the role and meaning of forests in American life and made some innovations in its program in that direction, it seemed immobilized by its commodity forest traditions. That condition of stalemate was encouraged by commodity forest advocates in the industry, by Congress, and by traditionalists within the agency.

As far as the agency was concerned, it was faced with new legislative mandates that required it to give some recognition to ecological objectives. This pattern of change and response had been the case in the past, and it continued to be so. On a number of occasions throughout its history, the agency's resistance to changes in forest management arising from the public had led to a reduction in its administrative authority, and its continued resistance to new directions urged by ecological forest advocates posed the same threat.

Because of the agency's rejection of national park objectives that disallowed commercial timber harvest, it had failed to acquire jurisdiction over such objectives and was unable to prevent establishment of a separate national park agency. As a result, it lost control over management of significant areas of the national forests, which became national parks. Later, because of its resistance to the permanent designation of forest areas as wilderness, it lost the authority to make the final decision about such proposals and Congress became the final arbiter. In the Multiple-Use Act of 1960, the agency had succeeded in retaining authority to decide the balance of uses designated in the law, even though it failed to retain ultimate authority over wilderness designations. Resistance to ecological forest objectives in later years led to the further loss of agency autonomy.

This further loss was marked by three pieces of legislation. The first, the National Environmental Policy Act of 1969, required all federal agencies to conduct an analysis of the environmental consequences of proposed actions, propose alternative courses of actions, and justify its decisions. The law thus did not require the agency to make specific policy choices in an environmental direction but only to identify their environmental implications and explain why it chose one course of action over the other. In this process, the agency was required to submit environmental impact statements to other agencies for their review.[33] The second law, the Endangered Species Act of 1973, was more substantive. It established species protection goals to which all public and private agencies under the supervision of the U.S. Fish and Wildlife Service must adhere. Species with population reductions sufficiently great that their continued existence was endangered or those that might reach that stage unless action was taken were identified as "threatened" and earmarked for action. The Forest Service, along with other agencies, was required to submit candidate species to the Fish and Wildlife Service for their "biological opinion" about such matters.[34]

The third legislative move that limited agency autonomy was actually two laws: the Clean Air and Clean Water Acts of 1970 and 1972. They and their subsequent amendments imposed more specific restraints on the

Forest Service. While both environmental programs arose in response primarily to urban circumstances, their mandate was slowly expanded to "keeping clean areas clean" rather than just "cleaning up dirty areas." In either situation, the Environmental Protection Agency had the authority to review agency actions. This gave rise to cases in which the EPA challenged standards applied by the Forest Service. For example, while the Forest Service sought to use BMPs as standards in protecting water quality from potentially polluting timber harvest practices, these standards were widely viewed as minimal. The EPA, when it became serious about non-point sources of water pollution, sought to require the Forest Service to accept the EPA's standard rather than the much weaker BMP standards. Several court cases sharpened the clash between agencies over these issues.

Debate over the National Forest Management Act of 1976 brought the Forest Service face to face with proposals that its decisions be subject to specific restraints, such as that timber harvest be limited to lands that were sufficiently productive to produce a new "crop" within a reasonable amount of time or that clear-cuts be prohibited.[35] The agency objected to such provisions on the grounds that they were matters of professional judgment that should be left to its staff, and for the most part it succeeded. However, the law provided enough requirements of a more general nature that ecological forestry advocates had some leverage from the statutory language to argue their case for revised policies.

One provision of the 1976 act that came to be of great importance in future years was the "diversity" clause, which required that the agency maintain species diversity in the forests. When this provision was incorporated into the regulations developed to implement the act, the agency faced the new requirement that it maintain the viability of a wide range of species, much beyond the threatened and endangered species, which were known as "sensitive species."

The 1976 act required that each national forest produce a plan for its management over the next fifteen years and that the public have an opportunity to submit comments on a draft plan before the final plan was approved. Under the NEPA requirements, the agency drew up its environmental impact analysis in conjunction with its plan. The two, in tandem, provided a considerable opportunity for the presentation of ecological objectives in competition with commodity objectives, and both citizens and scientists actively engaged in the process.[36]

There ensued an extensive critique of the agency's proposals. Over the course of time these critiques led to the preparation of citizen-originated plans as alternatives to the agency's, and they provided an opportunity for ecological objectives to be widely publicized even though they were not fully accepted by the agency. Citizen groups linked to many national forests arose to participate in this process, and as the agency developed regional plans for such forest areas as the Sierra Nevada in California and the southern Appalachians, so did the citizen organizations.

Ecological scientists played an important role in this process. The environmental analyses required for each plan provided a unique opportunity for their involvement. The law required that the analyses be interdisciplinary, and the process thus invited comment on the plans from scientists with far more extensive biological interests than the more narrow silvicultural and wood production experts who had been the dominant group of scientists in national forest affairs throughout its history.[37]

These scientists came from a wide range of disciplines in a variety of biological and hydrological departments in academic institutions across the country, and by their participation in national forest affairs via ecological forestry objectives, they served to fracture the agency's cohesive but limited perspective in commodity-based forest science. They played quite a different role in the national forest planning process than did citizen groups, and often they took the lead in identifying the implications of forest management for ecological objectives. But both groups helped to shape an ecological perspective that evolved in the last third of the twentieth century.

The appearance of a group of diverse biological and hydrological scientists in national forest affairs stands in marked contrast with the more traditional role of commodity-focused interests in forest science. The two factions were on separate paths. This new group of scientists participated in forest planning, contributed to the scientific aspects of policy issues, and advised citizen groups in preparing and evaluating forest plans. Ecological scientists gave the ecological approach to forest management a high degree of strength and continuity.

Facing an Active Public

Pinchot had shaped the Forest Service to operate mostly without the input of the public's advocates for wildlife, aesthetic, and watershed interests. The agency's slow and grudging acceptance of their objectives fostered the growth of an informed public outside the agency and made it difficult for the agency to collaborate with those citizen groups. A mutual sense of distrust arose and grew steadily after the 1960s, and the agency did not seem to be able to change that dynamic.[38]

Public distrust of the agency was reflected in ongoing criticism of its policies and projects. The agency expressed a similar distrust of the "environmental community." In an article in the *Journal of Forestry*, "Democratic Accountability in National Forest Planning," Nancy Manring summarized her view of the problem of "collaboration" between agency and public: "collaborative efforts may be constrained by what some agency officials have referred to as a longstanding paranoia—an absolute distrust of the public and environmental groups by line officers and most management."[39]

This public was far different from that of Pinchot's years with the agency. Citizen groups had become far more knowledgeable about the individual forests and could challenge the agency's site-specific actions on the ground. Scientists interested in a broad range of biological, archeological, and geological resources other than wood fiber worked in a wide range of institutions beyond the more limited traditional forest organizations. Citizens and scientists now generated their own forest plans to compete with those drawn up by staff within the Forest Service. Public education about forest ecology evolved to encompass a much broader range of forest resources and processes than the agency's focus. A wide range of handbooks, guidebooks, and source books made available to the public information about the national forests far beyond what the agency provided.

The agency's response to its new public was often plaintive: "the public does not understand us; we try to do a good job but are still subject to criticism." It was often manipulative, fostering changes in language—for example "vegetative management" for "logging" and "wood harvest"— rather than being straightforward in its analysis of issues. It was prone to denigrate criticism as simply extremist, pointing to street demonstrations rather than the quieter, more extensive participation by citizens and scientists in forest planning. It often relied on personal judgments about environmental and ecological forest circumstances while placing them in the context of firm knowledge.[40]

The Forest Service was rarely able to assign agency resources to the wider range of forest issues that attracted considerable public interest; neither could it relegate wood production to a subordinate role in management. Its still limited response to wildlife, watershed, and aesthetic objectives was symbolized by the relegation of much of their affairs to the little known and subordinate agency section known as WFW and dealing with watershed, fish, wildlife, air, rare plants, soil, and endangered species. The WFW Web site describes its primary role as research and explains that "information provided by this research is crucial to helping the agency comply with requirements of key environmental statutes, including the National Forest Management Act, Endangered Species Act, Clean Water and Clean Air Act." Thus, these responses to wildlife, watershed, and aesthetic issues were not management objectives but "compliance requirements."[41]

The Forest Service was most successful in turning back the influence of the public by restricting administrative appeals. As with many government administrative bodies, the agency had developed over the years a process by which those who felt aggrieved could appeal administrative decisions. In earlier years, most such cases involved the conditions of user permits, such as grazing or mining permits. But the new responsibilities and rules arising from the National Environmental Policy Act, the Endangered Species Act, and the National Forest Management Act gave rise to a host of new types of agency decisions that prompted appeals from those who felt aggrieved.

This made the appeal system the focal point of both citizen efforts to hold the agency to its responsibilities under the law and efforts on the part of those hostile to environmental and ecological objectives in forest management to curb citizen action. This opposition sought to limit the number of parties who could appeal, requiring them to submit extensive justifications for their case and restricting the time given for them to appeal and the time the agency had to respond with the frequent provision that if that time limit was not met then the agency decision would stand.

In 1992, Congress revised the appeal system but did so in a manner that retained most of the process. But with a Republican majority in Congress after 1994, support for more restrictions grew. Following the dramatic forest fires of the late twentieth and early twenty-first centuries, which led to accusations that the appeal process had hampered the agency's ability to deal effectively with them, the George W. Bush administration worked with the agency and the Republican majority in the Congress to greatly restrict use of the agency's appeal system. This was one of the more successful of the Forest Service efforts to turn back the environmental/ecological initiatives of the last third of the twentieth century.[42]

The Public Discovers the Forest Service Budget

Beyond their interest in the substantive direction of the Forest Service, more vocal and active segments of the public became interested in the continuing tussle between Congress and the agency over its finances. While in earlier years this drama had been carried out within the halls of Congress, it now took place in a more public arena that involved citizen environmental organizations, taxpayer groups, and action by the Congressional Budget Office (CBO).[43]

Some of these tussles involved rather traditional skirmishes between Congress and the agency. Increasingly, however, the agency emphasized the intangible benefits from its management amid pressures to specify these benefits with more quantitative accounting. And it adopted the practice of justifying a number of activities, especially road building, as "joint costs" that should be attributed not just to wood harvest but also to other areas, such as access for recreation and wildlife enthusiasts.[44]

An entirely new issue arose that brought greater public visibility to the agency's budget. Financial studies conducted by outside groups began to demonstrate that income from harvesting timber on most forests did not cover its costs; it was a financial loss to the nation and thus was not worth the cost.[45] The argument was first developed and publicized by several staff members of the Natural Resources Defense Council in 1980.[46] It stirred considerable public debate. The Wilderness Society, which had hired staff well versed in economic analysis under its new executive, William Turnage, reviewed the finances of a number of specific national forests and found that each reflected "below cost" returns from timber harvesting.[47] The appropriations committee of the House of Representatives commissioned a study from the Congressional Research Service (CRS), which confirmed the "below cost" results of the Forest Service harvest program. The General Accounting Office did an analysis of the problem as well.[48]

The CRS's accounting work on the agency's timber finances was conducted by Robert Wolf, a professional forester who became the nation's foremost authority on agency finances dating back to 1905; his work played a significant role in the "below cost" timber financing debates that ensued. In the larger world of citizen action, Randy O'Toole, the Eugene, Oregon, economist, became influential in supporting the financial critiques of the agency and served as an additional authoritative voice in these debates.[49]

By 1984, the public controversy over the finances of the Forest Service led to instructions from the House budget committee that the agency develop an accounting system that would pin down income and revenues through specific programs rather than gross indicators. The agency did so in a financial report, but at the same time it presented employment and program accounts that, somewhat in its traditional fashion, sought to divert the financial debate by emphasizing the jobs that the agency created for local communities.[50]

By 1988, the issue had become a central feature in the larger debates over national forest affairs. The Wilderness Society, the main economic analyst for citizen groups, was critical; timber industry groups complained of the way in which citizen groups interpreted the data; Randy O'Toole claimed that the agency designed the system "to justify uneconomic timber sales by the jobs they support."⁵¹ Robert Wolf objected that the agency was still failing to deal effectively with costs in the manner that standard accounting practices prescribed. The new accounting system, he wrote, "purposely understates the costs and presents a picture that justifies the money-losing timber program." In 1985, a task force of the Society of American Foresters wrote, "No accepted method yet exists for identifying and quantifying nonmarket benefits for comparative analysis, and adequate techniques do not exist for allocating joint costs to several different resource programs."⁵²

In his account of the history of the agency's financing, Robert Wolf described its accounting practices as the "legacy of Gifford Pinchot," stemming in a rather direct way from the first chief forester's claim that he could "make the forests pay." That mantra compelled the agency to live up to its claim in a convoluted and often even humorous fashion.

One might well emphasize these financial maneuverings as simply another example of the agency's inability to control its own affairs—in this case its financial practices. These now were as extensive a matter of public debate as were the issues of multiple use and ecological forestry. To improve the appearance of its accounting, the agency was often tempted to develop ad hoc and sometimes bizarre arguments, as when, for example, it extended the time period of road depreciation costs on the twelve forests in the Rocky Mountain region to 235 years.⁵³ By the early twenty-first century, the Forest Service could no longer insulate its financial accounting from wider public scrutiny as it had done with wood production objectives in earlier years.⁵⁴

The Struggle between New and Old

The innovations brought to national forest affairs by ecological forest advocates created an intensely hostile reaction on the part of wood production and grazing interests. Off-road vehicle advocates as well as the developed recreation industry (e.g., ski resorts) joined in. The wood production industries had long opposed wilderness designations as well as hiking trails and wild and scenic rivers designations because they reduced the amount of land available for timber harvest, and livestock interests had opposed the expansion of forage use on behalf of game wildlife. Ecological forest objectives now posed even greater threats, for they could not be dealt with just by segregating their use from the "general forest"; they required restrictions on how wood production activities and livestock grazing themselves were conducted.[55]

The substantive issues on which this opposition concentrated were the two main objectives of ecological forestry: species and habitat management on the one hand and watershed management on the other, and each of these brought a variety of challenges.

To counter species protection objectives, the opposition argued that biologists used faulty methods to count species and, in particular, that they considered subspecies to be distinct from main species when, in fact, if subspecies were added together, then the population was sufficiently large that it was not endangered. The critics' most well-known case was the spotted owl, which biologists divided into three regionally located species—northern, Californian, and Mexican—and considered each one separately to determine their population. The wood production industry argued, however, that they were all one species and, taken together, were sufficiently numerous that they were not endangered or threatened.

The industry objected to the Forest Service policy of providing for the "viability" of species, which extended species and habitat management to a much wider range of plants and animals. It was especially critical of the "survey and manage" policy, which required that species surveys be made in conjunction with harvest projects so as to provide data indicating the impact on species brought about by changes in habitat from wood harvesting.[56]

Both the wood production and the grazing industries objected strenuously to ecological policies intended to protect watersheds and especially urban water supplies and aquatic species. Their lawyers used arguments similar to those used for terrestrial species by persuading the courts (later overruled on appeal) that wild salmon and farm-raised salmon should be counted as one rather than separate species, and thus salmon populations were not threatened and did not require restrictions on headwater wood harvests to protect their aquatic habitat. They argued that state "best management practice" actions were sufficient to protect streams from siltation caused by logging and that Forest Service action beyond BMPs was not necessary.[57] When this dispute led to litigation, the federal court disagreed; the agency had to meet the EPA and not just the BMP standard.

When species/habitat and watershed issues came to a head in the planning and environmental analysis processes, the Forest Service often found that its lack of information, experience, and expertise in such matters made it vulnerable. How was it to keep track of species "viability"? It had no fund of information to enable it to do so. At first it worked out an approach in which it selected certain "management indicator species" (MIS), which supposedly would be representative of larger numbers of habitats and species and, if kept in a satisfactory condition, would suffice.[58]

Forest Service staff, trained in wood production and less than enthusiastic about species/habitat management, relied extensively on guesswork rather than data. It took up the argument that if sufficient habitat was available for broad classes of species, then it was not necessary to count the species themselves. It amounted to a management shortcut much like the one it had already developed for trees by shifting from counting trees on the ground to estimating their wood volume from aerial photographs. In cases involving the adverse effect of harvests on watersheds, it was prone to say simply that the effect would be minimal and that while harvest on one watershed would create more erosion, the protection of another would compensate for loss in the watershed as a whole.[59]

One could easily read these agency deficiencies as the result of longtime neglect of both wildlife and watershed issues, in contrast to the preoccupation with wood production that had enabled the Forest Service to build silvicultural expertise.

The opponents of environmental and ecological forestry tended to shift their arguments from substantive to procedural issues. The problem, they argued, was that environmental and ecological forestry advocates had too much leverage from using recently developed procedures. So they focused agency reforms on measures to limit the application of the environmental impact analysis, to limit the authority of the Fish and Wildlife Service to determine the status of species under the Endangered Species Act, to restrict the way in which the Wilderness Act could be used to expand the national wilderness system, to limit the use of appeals within the agency, and to restrict the use of the federal courts. While it was difficult to revise laws, it was possible to restrict their application.

Opposition to the environmental and ecological innovations in forest management gained considerable support after the 1970s from like-minded members of Congress and their close association with the Republican Party. This opposition group enjoyed extensive influence with the Reagan administration in the 1980s and from the George W. Bush administration after 2000. Until 1994, this influence had been held somewhat in check by a Democratic majority in Congress, but after the Republican congressional victories beginning in that year, this roadblock was weakened. When the Republicans gained control of both Congress and the White House in the election of 2000, the opponents of environmental forestry reached a new peak of influence that enabled them to reverse many of the new, though limited, environmental and ecological approaches that were being used in the Forest Service. The environmental/ecological opposition now controlled the crucial resource committees in both the Senate and the House, and the George W. Bush presidency brought that opposition into its administration, which resulted in a vigorous and effective reversal of the direction of national forest affairs.[60]

This reversal in policy by the Bush administration was reflected in several events that indicated the depth and scope of change. There were, for example, two contrasting statements by the last chief forester during the Clinton administration, Mike Dombeck, and the first of the Bush administration. Dombeck had outlined the agency's resource agenda, which stressed agency objectives for the future: watershed health and restoration, ecologically sustainable forest and grassland management, recreation and roads—all within a context of attempting to restore public confidence in the agency.[61]

Dombeck's successor, Dale Bosworth, however, spoke not of objectives but of threats: invasive species, unmanaged off-road vehicle use, loss of open space, and wildfire. Bosworth simply ignored ecological forest objectives, seemed little concerned about the public's loss of confidence in the agency, and spoke disparagingly of policies such as the protection of roadless areas, which Dombeck had advocated. On a more programmatic level, the change in the administration of forest policy was made clear in the agency's reversal of ecologically oriented planning proposals for the Alaskan forests, the Pacific Northwest forests, the Sierra Nevada, and the southern Appalachians as well as many individual forest plans.[62]

The most extensive example of the reversals that the new Bush administration brought about were changes in the regulations for forest administration that had been put in place by the National Forest Management Act of 1976. The Clinton administration had sought to bring into one system of regulations the various legislative mandates under which the agency worked, including the National Environmental Policy Act, the Endangered Species Act, and the National Forest Management Act. It had appointed a committee of scientists to set the direction of the regulations, which proceeded with the objective of fostering the "ecological sustainability" of the forests. The new directions included measures to foster environmental and ecological objectives throughout the national forest system, to manage on behalf of the viability of species, and to bring ecological science more fully into forest administration by providing for "peer reviews" by each national forest for the scientific content of its plans.

These proposals, which were formulated during the second term of the Clinton administration and issued toward its end, energized the environmental/ecological opposition and immediately came under fire. The presidential victory for the Republican Party in the 2000 elections and the George W. Bush administration gave them the opportunity to launch a counterattack in which the proposed regulations were withdrawn and revised considerably. Their main thrust was to give forest supervisors considerable freedom to make their own decisions about species and habitat, relieve them of conducting environmental reviews of forest plans, eliminate the requirement of obtaining "biological opinions" from the Fish and Wildlife Service for endangered species decisions, and provide the option of overruling scientific judgments when they felt policy considerations might warrant it.

The Bush administration regulations represented a far-reaching effort to use administrative discretion to reverse environmental and ecological objectives fostered by the public and scientists. The administration acted vigorously on behalf of the extractive industries and the motorized recreation industry and set the stage for a future of even more intense controversy in national forest affairs.

The Legacy of Ecological Forestry

The last third of the twentieth century brought to the Forest Service new circumstances that played a significant role in shaping how it faced the future. Two of these new circumstances—ecological and watershed objectives—involved substantive resource issues concerning the health of resources that had been compromised heavily by the past accommodation of resource users.

There were also new constituencies, one of citizen groups and the other of scientists, both inspired by an ecological perspective. Both sought to advise the agency about appropriate directions in both ecological and watershed management. They were far different from the longtime group of resource users who could be placated by more resource allocations for their use; these new advocates emphasized "healthy" forest resource conditions, which called for ecological science to play a significant role.

There were innovations in planning, with each national forest requiring a plan as directed in the National Forest Management Act of 1976 and NEPA, all of which brought to forest affairs opportunities to work out new substantive directions in forest management and establish a new setting in which resource advocates could advance their proposals.

The legacy of this period of innovations in national forest affairs was far more comprehensive than those of earlier periods and thus constituted a more complex set of challenges for the agency. However, greatly inhibiting the implementation of these resource objectives were the reforms in national forest affairs promoted by the George W. Bush administration. These reforms were much more than simply changes in policy direction. Those changes were intended to free the agency from the restraints legislative mandates had recently placed on its autonomy. Among these restraints were challenges from the Environmental Protection Agency and the U.S. Fish and Wildlife Service. For a number of years those agencies had been given the authority to review and question decisions made by the Forest Service. With respect to water quality and endangered species, agency supervisors were once again free to make their own judgments about such matters. At the same time, the reforms enabled agency supervisors to develop fifteen-year plans for their forests without conducting environmental analyses, a change that greatly limited the ability of citizens to subject the resource consequences of the plans to scrutiny, an intrusion the agency had long resented. Finally, regulations that had provided for peer review of scientific assessments were withdrawn. Supervisors could again decide for themselves just what conclusions should be drawn from the scientific knowledge available.

The U.S. Forest Service thus faced the twenty-first century amid conflicting impulses. On the one hand it had responded with reluctance to the multiple-use and ecological forest demands from society at large, and it was now hampered by the agency's inability to confront the changes in the meaning of forests in American society with a firm sense of direction. But it also faced the implications of the new regulations the George W. Bush administration had adopted in its attempt to free the agency from the restraints of citizens, scientists, and laws and regulations.

At this early stage of its second century, the U.S. Forest Service faced not only a new legacy of ecological objectives that scientists and citizens insisted the agency work into its long-standing silvicultural program but also the necessity that it do so amid a set of rules and regulations hostile to those objectives. These conflicting forces further complicated the relationships between the agency and the larger world of public affairs.

# EPILOGUE   Facing the Future

AS THE U.S. Forest Service stood at the threshold of its second century in 2005, it faced the future with a compelling inheritance. From the vantage point of those charged with shaping agency affairs in the past, this inheritance took the form of a legacy, a responsibility carried over from earlier years to its future programs. From the vantage point of future administrators, it constituted an institutionalized influence, often unwelcomed, inherited from the past.

In the previous chapters, I have traced the history of the Forest Service as an accumulated sequence of legacies: the silvicultural imperative of the early decades, the multiple-use legacy of the middle twentieth century, and the ecological forest legacy of the last third of that century. In this final chapter I explore briefly some implications of this inheritance and some of the ways in which it helped to shape the agency's responsibilities as it began its second century.

Gifford Pinchot's decision to limit the mission of the Forest Service to wood production hung heavily on the agency throughout the twentieth century. Timber production from the national forests reached a peak of 12 billion board feet during the Reagan administration and, due to a variety of factors, declined steadily thereafter to some 3 billion by the end of the agency's centennial. In 2005, Dale Bosworth, the agency chief, declared that no longer was wood production the dominant focus of the Forest Service; it was now replaced by recreation and forest restoration. Yet despite declining harvest levels, agency issues continued to reflect the controversial role of wood production in many an agency proposal. In the late twentieth and early twenty-first centuries, the relative influence of this silvicultural imperative waxed and waned, falling in Democratic presidential administrations and rising in Republican ones. Amid these fluctuations, Pinchot's wood production legacy to the agency remained.

One relevant example of this persistence during the George W. Bush administration involved the agency's management for the recently established Giant Sequoia National Monument in the Sequoia National Forest in California. That forest, established in 1908, contained a large grove of redwoods. Citizens had long sought greater protection for the trees; their action succeeded when the Clinton administration established the Giant Sequoia National Monument to protect them. It was to be administered by the Forest Service under a provision that none of the trees included in the monument area would be harvested for timber.

Yet the agency's management plan for the monument provided that 7.5 million board feet of redwoods per year be harvested on the pretext that this was essential to protect the trees from fire. When citizens brought suit against the Forest Service on the ground that the plan was illegal, the district judge agreed and stated that "the Forest Service's interest in harvesting timber had trampled the applicable laws." In its critique of the agency's proposal, citizen groups compared the plan unfavorably with the fire prevention program carried out by the National Park Service at the nearby Sequoia National Park and proposed that the monument be transferred to the Park Service for administration.[1]

Intimately associated with wood production was the forest road system, which provided the agency with an even more overwhelming burden. The national forests, largely western at their origin, were established in a heavily roadless region, and to market timber, the Forest Service became heavily involved in road building, a venture that grew steadily. By the end of the twentieth century, the Forest Service had built some 343,000 miles of road. During this time, little thought was given either to halting the pace of growth or to effectively maintaining the "system."

During the second half of the Clinton administration, Mike Dombeck, the chief forester, finally gave firm attention to the problem. The cost of building and maintaining forest roads was astronomical, and it did not make sense, Dombeck argued, to keep expanding the road net when construction in ever-higher elevations, with steep slopes and fragile soils, made their protection even more difficult and when an effective maintenance program lagged so far behind. His proposal was twofold: quit building roads in areas that had none and selectively phase out roads already built. The first proposal, although widely supported, faced vigorous opposition through the initiative of the agency during the George W. Bush administration; the second proposal was carried out more quietly and with some success, but it is difficult to determine just how much. As the agency neared its second century, it was saddled with an inherited road system that it hardly knew how to maintain or control.[2]

The legacy of the multiple-use program, codified in the Multiple-Use and Sustained-Yield Act of 1960, presented a more subtle but equally challenging future for the agency. When the Forest Service took up the formal task of adjusting the "forest uses" identified by the act, it did so largely in terms of ironing out differences between competing users. And it did so with the confidence that it would satisfy all of them by increasing benefits to all. Soon, however, that multiple use also involved "resource conditions"—issues of just how the increasing use of one resource could harm the use of another. Harvesting practices could threaten water quality, roads could disrupt the movement of wildlife, excessive deer populations could destroy vegetation, and mining could impact both water users who drew on underground supplies and valuable wildlife habitat.

To justify its decisions in the context of its multiple-use mandate, the agency was increasingly called upon to identify more precisely the impact of the use of one resource on another. This shifted the task of determining the "balance between uses" to technical "resource impact" information and the need for scientists to help implement multiple-use decisions. But what if the scientists on whom it relied came to conclusions that challenged policy decisions that agency administrators preferred? When that happened, there was inevitably friction between independent scientific judgment and policy decision making.

It was precisely this kind of situation that led agency employees to organize a group called Forest Service Employees for Environmental Ethics (FSEEE), which took up the defense of agency scientists in a number of controversies with administrators.[3] In turn, administrators sought to be free from such influence and to obtain authorization to overrule their own scientific advisers. By 2005, multiple use, in one of its many forms, had evolved to this technical stage to set the tone for how it would be worked out in future years.

The multiple-use mandate also called for the agency to modify its long-standing practice of concentrating its personnel in wood production and begin to employ other types of professionals such as landscape architects or wildlife biologists. Yet this came only slowly after the 1960 act, first with recreation and more slowly with wildlife and wilderness.[4] The agency lagged still further behind with respect to those professionals in the newer environmental subjects such as air and water quality, soil, or endangered species. The agency's limited response to these responsibilities was reflected in the formation of the quite subordinate division in its Washington office known as "WFW," with a topical subject matter identified as "watersheds, fish, wildlife, air, plants, soil, and endangered species." WFW activities were devoted largely, though not exclusively, to gathering and communicating the latest relevant scientific knowledge in response to a variety of new legal mandates. But these commitments were not yet very well incorporated into management responsibilities, and the agency entered its second century with severe limitations in the balance of its professional skills, a deficiency that the agency would carry for years to come.[5]

New management skills appeared in the Forest Service in more piecemeal form, frequently in cooperation with other agencies. An active air quality monitoring program in the northern Rockies, for example, was initiated by administrators of the Shoshone and Bridger-Teton National Forests to aid in implementing agency clean air responsibilities. It became a part of the Greater Yellowstone Area Clean Air Partnership.[6]

In another example, the Forest Service, through its eastern region, cooperated with federal, state, and private entities to improve water quality in the Chesapeake Bay. It joined with other organizations to form the group Forestry for the Bay to work out watershed programs to reduce pollution from streams entering the bay. Initially emphasizing the retention of streamside buffers, the program began to focus on pollution resulting from the conversion of forest land to development, and it established a regional goal of adding 695,000 acres of permanent forest land to the bay's watershed. This more selective development of agency management expertise seemed to mark one course of the agency's future.[7]

The emphasis on ecological forestry that emerged in the late years of the twentieth century presented a new challenge to the Forest Service and to forest management in general, one that was even more fundamental than the agency's task of working out a balance among multiple users. It involved accepting a far different conception of just what a forest was, one that incorporated both the range of relevant plant and animal species and a broad view of the forest ecosystem of which they were a part.

To the traditional forester, a forest was composed primarily of woody species, the most important of which were those of commercial value. Ecological forestry expanded this view of forest composition to hundreds and thousands of species of plants and animals, each of which was considered to play an integral role in the forest ecosystem. For traditional foresters, this expansive concept was incomprehensible. The difference was one of values arising from world views that seemed beyond "rational" debate. However, in the new twenty-first century, the "foresters" were nevertheless responsible for a biological forest system that was far more diverse and complex than they had thought during the previous century.[8]

In an earlier chapter I detailed the reluctance of the agency to devote the necessary resources to the elementary task of species inventory, including counting the populations of "management indicator species" and undertaking "survey and management" tasks. After a number of years of such work, it was curtailed, with the agency arguing that it had neither the skills nor the budget to continue. In revising planning instructions, the Bush administration dropped this responsibility and took up the argument that population surveys were not necessary if sufficient appropriate habitat was available.[9]

A classic example of species recovery, the Kirtland's warbler illustrates the widely acknowledged necessity of population counts as well as appropriate habitat in species management. That warbler, which is found primarily in the northern lower peninsula of Michigan, nests only in the levels of jack pine forest below twenty feet in height. Jack pines are regularly managed to produce the desired habitat. Yet habitat acres do not indicate species numbers, and to arrive at that figure, a count of warblers has been taken annually and taken as an indicator of the success of the restoration program. Those numbers rose from two hundred in 1991 to fourteen hundred in 2005. In order to fulfill its statutory mandates with respect to species management, the Forest Service still faces the need to count sensitive species in the years ahead as firmly as it inventories tree species.[10]

In the early years of the twenty-first century, during the presidential administration of George W. Bush, this sequence of legacies from the agency's past became even more tangled. Bush policy makers rejected an outright assault on the environmental/ecological legislation of the last third of the twentieth century, which they sought to undermine. They instead chose to use the more pragmatic strategies of administrative power to achieve their objectives. Through these devices they sought to reduce the influence of the public on its actions, to override the influence of science, and to reshape administratively the meaning of legislation already passed. The courts rejected many of these strategies as contrary to law, but these defeats seemed only to stimulate the administration to search for more effective ways to circumvent the statutes.[11]

This practice continued throughout the Bush administration so that as of August 8, 2008, the administration proposed regulations that considerably modified Forest Service implementation of the Endangered Species Act of 1973. That act required that all federal land management agencies refer potential endangered species issues to the Fish and Wildlife Service for a "biological opinion" in order to determine the status of species and, if endangered, recommend actions to halt their decline or restore their numbers. The new regulations dropped this requirement and provided that forest supervisors themselves could decide if an opinion was necessary, thereby shifting control over such matters from the wildlife specialists to the Forest Service, which had far more limited resources to conduct the relevant scientific studies.[12]

What legislation, court decisions, or regulations should forest supervisors follow? Amid attempts by the public to urge this or that course, which one would enable the agency to move forward? If forest supervisors were permitted to choose their own science, what effect would this have on the role of the inevitable contradictory scientific conclusions? The Bush architects of national forest policy had guaranteed quite a new legacy for the agency, one of continuing contradictions and increasingly more elaborate and intense controversies involving complicated relationships among legislative, executive, and judicial branches of government. All of the legacies of the past century now had to be filtered through the maze of "authorities" that the Bush administration had constructed. Old and new were now in more confused combat than ever before.[13]

In the coming second century of its history, the U.S. Forest Service will undoubtedly face what sociologist Robert Merton called the "unanticipated consequences of purposive social action." The agency will also have to cope with those consequences, but it will also carry into its second century a legacy of highly influential "directions" established in earlier years and that continue to color its policies in later years. Robert Wolf had referred to Gifford Pinchot's promise in 1905 that he could "make the forests pay" as a legacy that his successors struggled to live up to.[14] The later mandates of the Multiple-Use Act and the varied requirements of environmental and ecological legislation that the agency attempted, albeit reluctantly, to realize were also legacies that shadowed the agency in the new century.

This set of historical legacies and inheritances is the most significant element of the first century of Forest Service history. They pose the continuing challenges for its second.

 NOTES

Works cited here in shortened form (author, title only) are listed in full in the "Sources" section.

Preface

1. See the bibliography for a list of works about the Forest Service from which I have drawn material for this history. Several biographies also contain important information about the Forest Service. They include: Meine, *Aldo Leopold;* Rodgers, *Bernhard Eduard Fernow;* Drummond, *Enos Mills;* Morrison, *J. Horace McFarland;* Pinkett, *Gifford Pinchot;* and Miller, *Gifford Pinchot and the Making of Modern Environmentalism.*

2. The most recent biography of Pinchot does little to change this traditional version of Pinchot's early role in the development of the U.S. Forest Service. See Miller, *Gifford Pinchot and the Making of Modern Environmentalism.*

3. Three books that together cover much of the role of wood production in the Forest Service are Clary, *Timber and the Forest Service;* Hirt, *Conspiracy of Optimism;* and Langston, *Forest Dreams, Forest Nightmares.* See also Clepper, *Professional Forestry in the United States.* For wildlife in the late nineteenth century, see Reiger, *American Sportsmen and the Origins of Conservation;* see also Trefethen, *American Crusade for Wildlife,* for relevant pages on the wildlife in the national forests.

4. Good examples of the legislative/administrative history of the Forest Service are Steen, *U.S. Forest Service,* and LeMaster, *Decade of Change.*

5. The focus on agency choices as the historical data on which this book is based deserves repeated emphasis. The aim is to identify those choices in which the agency accepted some options and rejected others and to identify patterns in the extensive series of decisions made over the years. Legislative and judicial decisions, the more usual ingredients of agency history, I have relegated to the background; administrative choices are considered as the main body of evidence. While legislation and court decisions establish the framework of legal history, administrative choices reveal more fully the innumerable facets of relationships between the wider society and the agency.

6. The Forest Service DVD, *The Greatest Good,* is an example of a history that draws heavily on the impact of dramatic events on the reader/viewer, a technique that obscures much of the history lying behind those events.

CHAPTER 1. A Century of Change

1. The role of government as a "last resort" caretaker was played with particular frequency in New York, Pennsylvania, and the Great Lakes states, where abandoned lands became publicly owned and then were organized into state- or county-owned public forest lands.

2. Another example of government serving as the property owner of "last resort" would be the drylands of the plains states, many of them subject to wind erosion in the 1930s and quite marginal in their potential productivity. In this erosion control program, those who unsuccessfully attempted to eke out a living on submarginal plains land were "resettled" elsewhere, and their lands would then be purchased and managed by federal agencies. More than 4 million acres of such lands were purchased in this manner and eventually were administered by the U.S. Forest Service as "national grasslands." See Moul, *National Grasslands*. See also Forest Service Web site, "The National Grasslands Story."

3. The application of forest homestead legislation to the Coronado National Forest in southeastern Arizona is described in Hadley, "Grazing the Southwest Borderlands," esp. 108–10.

4. See Clepper, *Professional Forestry in the United States,* esp. 55–68 in chap. 5, which is titled "The Forest Service: Its Struggle for Survival" and which covers the years after the Pinchot-Ballinger controversy. There were many attempts by ranchers, water power interests, and speculators to take advantage of the situation and abolish the Forest Service. It was not until 1920 that the agency "commenced to get its heads up." Ibid., 57.

5. William D. Rowley, in *U.S. Forest Service Grazing and Rangelands,* describes the ongoing attempts by stockmen to obtain private rights to rangelands. They were aided by Senator Robert N. Stanfield (R-Oregon), a stockman who introduced a bill recognizing the vested rights of stockmen in the grazing lands for which they held a permit. Throughout the West in the spring and fall of 1925, he held hearings that served as opportunities for continued attacks on the Forest Service. In response the next year, William B. Greeley, chief of the Forest Service, issued "New Grazing Regulations on National Forests," which granted a number of significant concessions to the stockmen. According to Rowley, "It appeared that Greeley had caved in to most of the demands of the industry in return for the recognition of the legitimate retention and administration of forest grazing lands by the Forest Service." Ibid., 136; see also 132–36.

6. See Jonathan Lash, *A Season of Spoils: The Reagan Administration*'s *Attack on the Environment* (New York: Pantheon Books, 1984); Ron Arnold, *At the Eye of the Storm: Caught in the Conflict* (Eugene, OR, 1984); James Watt, *The Courage of a Conservative* (New York: Simon and Schuster, 1985).

7. In national forest affairs, the research center's most well-known collaborator is Roger Sedjo, whose paper, "The National Forests: For Whom and for What," was published in the PERC policy series. See the PERC Web site for a comprehensive view of its views and the Sedjo paper.

8. The emergence in the West of significant environmental support for western public lands is the theme of Samuel P. Hays, "The New Environmental West," *Journal of Policy History* 3, no. 3 (1991): 223–48.

9. For the Sagebrush Rebellion, see Mary Reeves Bell, *The Sagebrush Rebellion* (Minneapolis : Bethany House, 1999); R. McGreggor Cawley, *Federal Land, Western Anger: The Sagebrush Rebellion and Environmental Politics* (Lawrence: University Press of Kansas, 1993); William L. Graf, *Wilderness Preservation and the Sagebrush Rebellions* (Savage, MD: Rowman & Littlefield, 1990).

10. For accounts of both proposals, see copies of newspaper clips, primarily from throughout the West in author file of the Hays Collection, Environmental Archives, Archives of Industrial Society, Archival Service Center, University of Pittsburgh (hereafter cited as Hays Collection, EA).

11. The continuing public support for publicly owned lands is reflected in an article by T. R. Reid, writing in the *Washington Post,* Sept. 21, 2005, under the headline "This Land Is Your Land, and It Needs You." Reid wrote, "Under the Federal Lands Policy and Management Act of 1976, Congress expressly promised 'to retain federal lands in public ownership.' There is no discernible movement in contemporary politics to reverse that policy."

12. For a description of the "regulated forest" and the objective of substituting it for the "ancient" forest, see Clepper, *Professional Forestry in the United States,* 108; Langston, *Forest Dreams, Forest Nightmares,* 103. See also Rodgers, *Bernhard Eduard Fernow,* 303, where it is explained that silvicultural management of the Forest Service tract at Cornell University was "to replace the old decrepit natural forest by a new, more valuable forest."

13. As early as 1909, Gifford Pinchot had expressed skepticism about the ability of the private forest industry to undertake sustained-yield forestry. See Pinchot, "Forestry on Private Lands," *Conservation of Natural Resource: The Annals of the American Academy of Political and Social Science* 33 (May 1909): 8. A variety of factors held private industry back, wrote Pinchot, primarily because the limited profitability in long-term, sustained-yield management was not sufficiently attractive to potential investors. See Clepper, *Professional Forestry in the United States,* 137.

14. When Pinchot ended his term as governor of Pennsylvania in January 1927, he resumed his conservation activity, including "vigorous agitation for federal regulation." See Clepper, *Professional Forestry in the United States,* 142–45.

15. Pinchot resigned from the American Forestry Association because of its failure to support the proposal for federal control of private timber cutting. In a letter to the

president of the association, Pinchot alleged that it "had fallen under the influence of the lumber interests." Quoted in Clepper, *Professional Forestry in the United States*, 158n18.

16. Clepper covers the regulation controversy in *Professional Forestry in the United States*, chaps. 10 and 11. Chapter 10 pertains to the "timber famine" background of the drive for regulation and chapter 11, to the ensuing controversy over regulation proposals.

17. Clepper traces the advocacy of regulation by Forest Service leaders from 1920 to 1950 in *Professional Forestry in the United States*, chaps. 10 and 11. For the demise of the regulation movement, see ibid., 162–63.

18. Henry Clepper writes that one result of the controversy over federal regulation of private forests was the growth of cooperation in fire and reforestation programs organized through the states and implemented through state forest agencies. Clepper, *Professional Forestry in the United States*, 95–97. One focal point of this growth process came about during the proceedings on the Clarke-McNary Act of 1924, which provided federal grants to states. As originally drawn up by Rep. John D. Clarke of New York, the legislation provided aid for both regulation of cutting practices and fire control, but the bill was revised along the way to omit references to regulation and to concentrate on fire control. See Steen, *U.S. Forest Service*, 188. The provisions of the legislation tended to mute the antagonisms generated by the controversy at the national level.

19. For the growth of federal funding for state forest programs, especially in forest fire programs, see Steen, *U. S. Forest Service*, 185–95.

20. The wartime trend toward greater production from the national forests, leading to a focus on "maximum production," is charted in Hirt, *Conspiracy of Optimism*, 56–57. In the agency's annual report for 1948, the chief of the Forest Service wrote, "In handling the national forest timber resource the Forest Service is working toward intensive management for maximum continuous production." Quoted in ibid., 57. Ira J. Mason, head of the agency's Division of Timber Management, took the lead in pushing each of the regional offices to promote accelerated harvests.

21. Agency foresters were told that the forest was to be managed in the future based on certain assumptions about harvest volume and that these assumptions would enable the immediate allowable cut to be much greater. The assumptions were that fire suppression could eliminate 90 percent of the "natural fire loss," that pesticide use would reduce insect and disease loss by another 90 percent, that cut-over areas would be immediately replaced with superior seed stock, that new technology would permit harvest in sixty rather than one hundred years, and that careful logging would increase utilization and reduce waste. Negative effects on other uses would be mitigated. See Hirt, *Conspiracy of Optimism*, 57.

22. Clepper, *Professional Forestry in the United States*, 135–51, traces the idea of a "timber famine" in a chapter titled "Timber Famine Warnings: Prelude to Regulation."

23. See Clepper, *Professional Forestry in the United States,* 197–286, for a description of developments leading to "industrial forestry" in private industry.

24. For a series of articles on the Plum Creek proposals, see items by Phyllis Austin in *Maine Environmental News,* especially "Plum Creek's Big Plan," Feb. 10, 2005.

25. The chief of the Forest Service announced in summer 2003 that his agenda for the agency involved addressing the four threats: invasive species, unmanaged off-road vehicle use, loss of open space, and wildlife. He announced these objectives in speeches throughout the West. See, for example, the editorial "Forests Face Fresh Threats," *Denver Post,* Sept. 28, 2003. See also *Transition in Agency Publics,* by Michael Milstein, writer for the *Oregonian.*

26. In the early years of the Forest Service, Pinchot had excluded wildlife and aesthetic interests from the agency's program. After World War I, these issues, with the addition of outdoor recreation, steadily increased in significance in national forest affairs and were finally officially recognized in the Multiple-Use and Sustained-Yield Act of 1960. That theme is described in greater detail in chapter 3, "Evolution of an Agency Clientele 1920–1975," in this volume.

27. The committee of the National Academy of Sciences that drew up suggestions for reserve management proposed that portions of the Rainier Forest Reserve and the Grand Canyon Forest Reserve should be set aside as national parks. See Clepper, *Professional Forestry in the United States,* 27.

28. John Reiger makes an excellent case for the significant role of sports enthusiasts and publications in the run-up to the 1891 legislation authorizing the reserves and the executive action to establish them. See Reiger, *American Sportsmen and the Origins of Conservation,* 105–25. The sportsmen's role is also described in Trefethen, *American Crusade for Wildlife,* chap. 8, "Saving America's Forests."

29. For the prevailing view in southern California regarding the relationship between forests and water supplies, see Lockmann, *Guarding the Forests of Southern California.* The close connection between forestry and irrigation was reflected in the title of the American Forestry Association's publication, *Forestry and Irrigation.* That this connection was later severed was evident when that title was dropped.

30. Clepper, *Professional Forestry in the United States,* 62, describes the outcome of different approaches to wildlife taken by wildlife advocates, by Pinchot, and by President Theodore Roosevelt.

31. Clepper describes the limited agenda of the Governors' Conservation Conference in *Professional Forestry in the United States,* 48–53. He notes especially the absence of wildlife, aesthetic, and watershed issues and the preference for economic development.

32. The growth in the numbers of sportsmen is described in Trefethen, *American Crusade for Wildlife,* 175.

33. This transformation, from viewing wildlife as game to viewing it as an aspect

of biodiversity, which encompasses the full range of flora and fauna, is recounted in chap. 4, "Confronting the Ecological Forest, 1976–2005," in this volume.

34. Documents pertaining to the work of the Clinton administration's Science Advisory Committee are on its Web site "Committee of Scientists."

35. The agency's centennial DVD directly attributed public opposition to street demonstrators and forest-occupying "hippies" rather than to those who, in a far more quiet way, had mastered the system and employed the agency's own administrative decision-making machinery to achieve their objectives.

36. For a history of the forestry profession, see Clepper, *Professional Forestry in the United States*. The Forest History Society has pursued several projects that develop different aspects of the history of forest science. Among them are Steen, *Forest Service Research;* R. Keith Arnold, M. B. Dickerman, and Robert Buckman, *View from the Top: Forest Service Research* (Durham, NC: Forest History Society, 1994); and Harold K. Steen, *An Interview with Carl E. Ostrom* (Durham, NC: Forest History Society, 1994).

37. William M. Harlow and Ellwood S. Harrar, *Textbook of Dendrology Covering the Important Forest Trees of the United States and Canada* (New York: McGraw-Hill, 1937). Using the Blue Mountains of eastern Oregon as an example, Nancy Langston describes the process by which a broad interest in forest botany changed slowly to a focus on only a small number of plants: "Inspectors listed every plant species they could identify. In contrast, by the late 1910s when the timber sale business had drastically increased in the Blues, the silvics reports mentioned only trees valuable for timber." Langston, *Forest Dreams, Forest Nightmares*, 93. Gifford Pinchot insisted that "forestry" was not "botany," which seemed to steer his "scientific" focus to a selected group of commercially valuable trees. See Miller, *Gifford Pinchot and the Making of Modern Environmentalism*, 131–32.

38. The documents pertaining to Leon Minckler's activities are in the Leon Minckler file, Box 22, Forestry List 1, Hays Collection, EA. They include the proposal, the relevant correspondence concerning it, his description of the entire SAF episode in "A Report to the Forest Ecology Working Group, S.A.F., Sept. 17, 1974," his later "Proposal for an Environmental Quality Subgroup" (also rejected), and copies of his reports published by the National Parks and Conservation Association. A copy of his 1974 report is in Hays, *Forest Papers*,1:73–76, and a copy is in the Hays Collection, EA.

39. For example, Forest Service chief Edward Cliff (serving 1962 to 1972), once stated, "In recent years it has become quite obvious that 'near-natural' management is practically equivalent to no management at all. . . . It follows that it is not logical to authorize the establishment of virtually unmanaged areas [i.e., wilderness] of up to 100,000 acres." Quoted in Clary, *Timber and the Forest Service*, 170.

40. See Steen, *Forest Service Research*, 29n.

41. Clepper, *Professional Forestry in the United States*, 127. David Clary reports

that these twenty schools retained a heavy wood production curriculum. Clary, *Timber and the Forest Service*, 152. Paul Herbert, of the National Wildlife Association, reported several decades later that of twenty-seven forestry schools at that time, only one required a course in water conservation, three in recreation, and eleven in wildlife management. The advent of wildlife curricula reflects the distinctive state context of modern wildlife management and its contribution to the larger "conservation of natural resources" orientation.

42. Clepper describes the growth of professional forestry education in *Professional Forestry in the United States*, 123–34, and outlines the "desired curriculum" developed by the group headed by Henry Graves. In 1932, Yale University Press published *Forest Education*, written by Graves and Cedric Guise of Cornell University.

43. Federal land agencies jointly sponsored the Arthur Carhart Wilderness Training Center and the Aldo Leopold Wilderness Research Institute, both of which stressed training for wilderness management and both of which were located on the campus of the University of Montana at Missoula.

44. The Wildlife Society was among the many organizations that reflected the trend of growing public interest in wildlife. The work of the society as an organization of wildlife professionals can be followed on its Web site under the heading of the "Wildlife Society."

45. The growth of a wider interest in forest biology on the part of scientists in disciplines beyond silviculture is covered in chapter 4, "Confronting the Ecological Forest," in this volume.

46. The centennial video production, *The Greatest Good*, includes a segment in which Jeff DeBonis, founder of the Forest Service Employees for Environmental Ethics, briefly referred to the role of the "ologists" in the agency. The larger network of professional connections resulting from the influx of new specialists is reflected rather dramatically in the Forest Service's joint scientific reports, which by the late twentieth century often brought together specialists from varied federal and state agencies along with those in academic departments and some private organizations.

47. "Scientists Say National Forests Need Protection, No Logging," Apr. 16, 2002, in "Scientists National Forest Letter" folder, National Forests, Hays Collection, EA.

48. "Joint Letter to President Bush Regarding Management of National Forests," Apr. 30, 2002, and "SAF Urges President Bush to Support Scientific Forest Management," both in "Scientists National Forest Letter" folder, National Forests, Hays Collection, EA.

49. The most comprehensive book on the history of forest fires is Pyne, *Fire in America*.

50. For these developments in forest fire history see Pyne, *Fire in America*.

51. Ibid.

52. The Biltmore estate in North Carolina that Pinchot managed for a brief time

was the first such forest management program in the country and a place known as the "cradle of forestry" in America. See Harold A. Steen, ed., *The Conservation Diaries of Gifford Pinchot* (Durham, NC: Forest History Society and Pinchot Institute for Conservation, 2001), 48–61. For his view of the Chippewa Forest Reserve in Minnesota, see Pinchot, *Breaking New Ground*, 203–12.

53. For a discussion of more recent fire issues and the controversy over where related Forest Service funds should be spent, see Hays, *Wars in the Woods*, 174–75, 256–57.

54. For the "use book," see *The Use of the National Forest Reserves: Regulations and Instructions* (Washington, DC: U.S. Forest Service, 1905).

55. Forest Service staff labeled the influx of these larger recreational vehicles a "menace." See E. F. Meinecke to his superior, Dr. Haven Metcalf, Apr. 5, 1935, on the Forest History Society Web site.

56. See Paul R. Josephson, *Motorized Obsessions: Life, Liberty, and the Small-Bore Engine* (Baltimore: Johns Hopkins University Press, 2007).

57. Unmanaged ORV use was one of the "four threats" identified by Chief Forester Bosworth.

58. For agency action concerning both wildlife and watershed objectives, see chapter 4, "Confronting the Ecological Forest, 1976–2005," in this volume.

59. The WFW section was created by the Forest Service in the late 1990s to house the scientific and technical aspects of the agency environmental/ecological activity, with responsibilities specifically for "watersheds, fish, wildlife, air, plants, soil, and endangered species."

60. "Adaptive management" was used to describe a new form of applied knowledge in which a resource management agreement, recognizing the limited availability of desired information, stipulated that monitoring which would detail the results of projects would be an integral part of a management program and that management would be modified in light of that knowledge. Adaptive management and its implications can be understood more fully through the Web site of the Collaborative Adaptive Management Network and the Adaptive Management Blog Archive.

61. One example of a concession that forest users obtained was a "no surprises" policy with respect to management of endangered species on private lands. This agreement required landowners to take certain measures to protect selected species in return for a long-term guarantee (usually fifty years) that the landowner would be exempt from undertaking further protective action even if scientists uncovered new scientific knowledge about species on the lands in question. For a brief description of the policy, see Malinda C. Taylor, "Moving Away from Command and Control: The Evolution of Incentives to Conserve Endangered Species on Private Lands," in *Biodiversity Conservation Handbook: State, Local, and Private Protection of Biological Diversity*, ed. Robert B. McKinstry Jr., Coreen M. Ripp, and Emily Lisy (Washington, DC: Environmental Law Institute, 2006), 441–56.

CHAPTER 2. The Silvicultural Imperative, 1891–1920

1. For examples of such accounts, see Steen, *U.S. Forest Service,* and LeMaster, *Decade of Change.*

2. Herbert Kaufman's *The Forest Ranger: A Study in Administrative Behavior* is an excellent snapshot of the cultural ethos generated by the agency's sharp focus on wood production.

3. The extensive role played by sportsmen and their organizations in creating a climate favorable to establishing the reserves is well documented in Reiger, *American Sportsmen and the Origins of Conservation.*

4. See Lockmann, *Guarding the Forests of Southern California,* chap. 1, n. 26.

5. See Drummond, *Enos Mills.*

6. A committee was appointed by the New York legislature to report on the condition of the Adirondack lands; it was headed by Harvard botanist Charles S. Sargent. Early in 1885, it filed its report, which stated that the exploitation of the area was reducing "this region to an unproductive and dangerous desert." Graham, *Adirondack Park,* 104.

7. Quoted in Schneider, *Adirondacks,* 224.

8. Ibid., 225. The Adirondacks have spawned a considerable number of books. Two that might serve as a beginning point for the interested reader are Schneider, *Adirondacks,* and Graham, *Adirondack Park.*.

9. The views of Fernow and Pinchot about the Adirondacks can be followed in Rodgers, *Bernhard Eduard Fernow,* and Pinchot, *Breaking New Ground.*

10. The standard work on grazing in the history of the national forests is Rowley, *U.S. Forest Service Grazing and Rangelands.*

11. Clepper, *Professional Forestry in the United States,* chap. 5, "Battles with the Stockmen," covers grazing issues prior to World War I.

12. The report of the National Academy of Sciences (S. Doc. 57, 55th Cong., 2nd sess. [1898]) outlining proposals for reserve management that led to the Forest Management Act of 1897 agreed fully with the argument that grazing sheep caused severe injury to the grazing lands. See Clepper, *Professional Forestry in the United States,* 74, for quotations from the report.

13. For the role of Albert Potter, see Rowley, *U.S. Grazing and Rangelands;* Clepper, *Professional Forestry in the United States;* Steen, *U.S. Forest Service;* and Hays, *Conservation and the Gospel of Efficiency,* 57–58.

14. See Clepper, *Professional Forestry in the United States,* 78–81.

15. Early Forest Service range work seemed to increase its reputation with stockmen. Potter sent questionnaires to stockmen using the range and found 75 percent favored the agency's permit program. An example would be the cooperative range management agreements according to which the stockmen would supply fence posts and the Forest Service would supply wire and staples. See Steen, *U.S. Forest Service,*

163. Steen also reports that the National Woolgrowers Association endorsed the program of the Forest Service, noting that "general policies were acceptable, although more to cattlemen than to sheepmen." Ibid., 163–64.

16. Langston, *Forest Dreams, Forest Nightmares*, 86–113, provides a cogent picture of the early work of Forest Service personnel as they began to apply their ideas about forestry to the Blue Mountain region of eastern Oregon. Meine, *Aldo Leopold*, 87–123, gives a similar account of national forest work by chronicling the early professional life of Aldo Leopold.

17. See Clary, *Timber and the Forest Service*, 123.

18. Filibert Roth administered the reserves while they were under the control of the Department of the Interior. Roth quoted in Rodgers, *Bernhard Eduard Fernow*, 294.

19. Pinchot quoted in Rodgers, *Bernhard Eduard Fernow*, 294.

20. "The retirement of Potter in 1920 signaled an end to the power within the Forest Service of grazing interests." This decline was marked also by "the advance of more conservative forestry practices that saw great danger in ever-increasing numbers of animals in the forests." Rowley, *U.S. Forest Service Grazing and Rangelands*, 120–21.

21. See George Perkins Marsh, *Man and Nature; or, Physical Geography as Modified by Human Action* (Seattle: University of Washington Press, 2003). Marsh's book is frequently referred to simply as *Man and Nature* but must be understood as focusing largely on physical geography, with an emphasis on forests and waters, and not on biology or recreation; it is thus especially relevant for issues pertaining to forest denudation, soils, and watershed effects rather than other human activities in forests.

22. For Pennsylvania's interest in watersheds as the basis of its public forestry program, see DeCoster, *Legacy of Penn's Woods*.

23. The phrase "favorable conditions of streamflow" reflects an eastern interest in water power for small industry and in adequate stream levels for navigation and floating logs to lumber mills. Such a phrase in a western context would stress the issue of sedimentation and its effect on reducing the capacity of irrigation ditches and reservoirs.

24. Quoted in Rodgers, *Bernhard Eduard Fernow*, 560. Bernhard Fernow, the nation's leading silviculturalist prior to the activity of Gifford Pinchot, was a forceful voice in highlighting the connection between forest cover and streamflow. He and Charles Kendall Adams, professor of political economy at the University of Michigan, jointly sought to refute the argument by lumbermen that there was no such connection, and Fernow sought to bolster the argument by citing measurements in Germany dating back to 1868. Ibid., 108, 109. However, in 1916, Fernow spoke of the earlier arguments as "overstatements" and defended them on the grounds of political strategy. Ibid., 560.

25. For the debate over the effects of sheep grazing on headwater soil erosion, see Rowley, *U.S. Forest Service Grazing and Rangelands*, chap. 2, "Pioneers in Grazing Regulation," esp. 27–29.

26. For the NAS report on grazing and watersheds, see note 12 above.

27. For the Forest Service research on the effects of clear-cutting on watersheds see Rodgers, *Bernhard Eduard Fernow*, 413. Three experiment stations had been established in Colorado and Arizona to study forest influences, including the effects of forests on streamflow.

28. According to Nancy Langston, in the Blue Mountains of eastern Oregon the role of watersheds faded into the background in the face of pressures for wood production: "Production of a crop was the emphasis here; concern about the watersheds—and any other aspect of the forest—vanished when the Forest Service set out to calculate annual cuts. When foresters tried to plan for the landscape as a whole, they focused on one small fraction: the timber outputs." Langston, *Forest Dreams, Forest Nightmares*, 164.

29. For standard treatments of wildlife see Reiger, *American Sportsmen and the Origins of Conservation*, and Trefethen, *American Crusade for Wildlife*.

30. See Reiger, *American Sportsmen and the Origins of Conservation*, chap. 5, "Development of the National Park Concept," on the support for Yellowstone National Park as a wildlife refuge, especially for bison.

31. Quoted in Clepper, *Professional Forestry in the United States*, 62. See also Reiger, *American Sportsmen and the Origins of Conservation*, 169-70, on how the Yellowstone National Timberland Reserve was established as the first forest reserve under the 1891 authorization act.

32. Quoted in Clepper, *Professional Forestry in the United States*, 62. For the negotiations between Pinchot and Roosevelt concerning the possibility of wildlife refuges in the national forests, see ibid. Clepper continues, "Pinchot was able to convince Roosevelt of the impracticality of such a measure, countering that under proper forest (including wildlife) management, with hunting permitted, game habitat and populations would improve." Ibid. Theodore Catton and Lisa Mighetto attribute Pinchot's opposition to wildlife refuges to his desire to appease ranchers: "To appease ranchers, he also opposed the Boone and Crockett's plan to set aside game reserves within the forests." Catton and Mighetto, *Fish and Wildlife Job on the National Forests*, 30. For this view, they cite Terry L. West, *Centennial Mini-Histories of the Forest Service* (USDA Forest Service, 1992), 38.

33. Trefethen, *American Crusade for Wildlife*, 196-201, covers the issue of deer on the Kaibab Plateau in the Grand Canyon National Game Preserve, 196-201.

34. In his early work on the national forests of the Southwest, Aldo Leopold devoted considerable time to game issues and worked on a plan for game refuges, "despite the fact that the Forest Service had as yet no congressional authorization to establish such refuges." See Meine, *Aldo Leopold*, 145 (quotation), 146, 154.

35. Meine, *Aldo Leopold*, 154. A bill to establish a national system of wildlife refuges, the Chamberlain-Hayden bill, was introduced, but it failed to gain support.

36. Quoted in Meine, *Aldo Leopold*, 135.

37. William Rowley reports that stockmen looked upon game as a threat to their forage, and "by 1918 some urban sporting groups called for federal laws to protect game in the national forests." Rowley, *U.S. Forest Service Grazing and Rangelands*, 77–78.

38. For the location of agency wildlife activities in the Division of Grazing until 1936, when the separate division dealing with wildlife was established, see chap. 3, "Evolution of an Agency Clientele, 1920–1975" this volume.

39. For a statement about the role of the "regulated forest" in early agency history, see Clepper, *Professional Forestry in the United States*, 108, which includes a succinct summary of the overwhelming influence of the strategy in the plan to liquidate the older, "unproductive" forests quickly in order to establish a new forest "from scratch." The new regulated forest would be more highly productive and guarantee the nation's future wood supply. See also Langston, *Forest Dreams, Forest Nightmares*, 103, 108–13.

40. Clepper, *Professional Forestry in the United States*, 108.

41. Langston describes a variety of ways in which markets tended to shape the specific provisions of timber contracts. She sums up the process, writing that "no matter what, forestry had to pay" and that "many matters of theoretical forestry were possible that the market conditions did not justify." Langston, *Forest Dreams, Forest Nightmares*, 96.

42. Langston describes this method of organizing timber harvests in concert with the mills. The initial step came in the formation of "working circles," areas of land whose boundaries were defined by the market for their timber. "A working circle," she writes, "usually included all the timber that would feed into a single large mill." Langston, *Forest Dreams, Forest Nightmares*, 164. This was a major step on the path toward defining timber areas not in terms of the requirements of sustained-yield management but in terms of mill-supply-oriented management.

43. The agency often increased harvests beyond sustained-yield requirements to keep mills running at full capacity. When private sources of timber on which the mills depended were exhausted, this pressure to increase the cut beyond silvicultural requirements was especially intense. See Langston, *Forest Dreams, Forest Nightmares*, 192.

44. The initial focus on "sustained yield" as a silvicultural standard had evolved into a focus on the role of the forests in sustaining the towns and jobs that were economically dependent on them. Thus, a former agency director of timber operations, Ira Mason, later a private timber consultant, argued that "sustained-yield forest management consists for a given forest in limiting the average annual cut to the continuous production capacity." Quoted in Langston, *Forest Dreams, Forest Nightmares*, 195.

45. The use of timber to finance transportation dates back to the earliest days of the Forest Service. Langston describes the way in which a timber contract in eastern Oregon was designed so as to enable the purchaser of the timber to use proceeds from the sale to finance a railroad to transport the logs to the mill. To the Forest Service, this

seemed to be the only way to get to market the trees that it wished to harvest. Langston describes a variant of this process in which forest management was pressured to revise a contract to cut the pine over six years rather than the original twenty so that the Hines Lumber Company could meet their debts acquired in building an access railroad. See Langston, *Forest Dreams, Forest Nightmares*, 182.

46. A brief statement about the marginal lands issue is in Hirt, *Conspiracy of Optimism*, xl. A marginal lands provision, which would confine wood production to lands that were similar in productivity to those that were attractive to private investors, was of "first priority" in the strategy of the environmental contingent debating the National Forest Management Act of 1976. See an account of the legislative proceedings by the leader of the environmental initiative, Brock Evans, "Shaping the National Forest Management Act of 1976: A Participant's View," reprinted in Hays, *Forest Papers*, 1:83–95 (copy in the Hays Collection, EA).

The limited wood production of marginal lands in the national forests was continually emphasized by Marion Clawson, a longtime member of Resources for the Future, a Washington, DC-based resource economics think tank. Clawson argued that the agency devoted too many resources "to improve forests on marginal lands where the returns on investments were low" and that they should be spent on "high-productivity" sites. See an account of Clawson's work by colleague Roger Sedjo, "Marion Clawson and American Forests" in Sedjo, ed., *Vision for the Forest Service*, chap. 1.

47. In his observations about New York forest management in the Adirondacks, Fernow associated forest aesthetics with parks and referred to them as a "luxury forest for the rich." Rodgers, *Bernhard Eduard Fernow*, 467.

48. Rodgers, *Bernhard Eduard Fernow*, 299.

49. For actions by the agency to apply aesthetic objectives to selected areas of forest management, see chap. 3, "Evolution of an Agency Clientele, 1920–1975," this volume.

50. Instructive statements about the national forests and mixing wood production and aesthetic objectives are in Drummond, *Enos Mills*.

51. The states that manage forests and parks under a common administrative umbrella are Massachusetts, Connecticut, New Jersey, Maryland, and Michigan.

52. For reference to the NAS recommendation of new national parks, Clepper, *Professional Forestry in the United States*, 27.

53. Pinchot and successor chiefs of the Forest Service persistently sought to transfer administration of both the parks and the forest reserves to the Department of Agriculture. That move was advocated in 1903 by President Roosevelt's Keep Commission on government efficiency and was proposed in a house bill introduced through the influence of Pinchot but successfully resisted by Representative Lacey. It continued to be proposed by Henry Graves, Pinchot's successor as chief of the Forest Service, as a way of "resolving" controversies arising from separate administrations.

54. When Mrs. Lovell White, chair of the Outdoor Art League of California, asked Pinchot to intervene with President Roosevelt and Speaker of the House Joseph Cannon on behalf of the proposed Calaveras Big Trees reservations, Pinchot replied through an aide, "Mr Pinchot thinks that in view of his official position he *cannot* with propriety actively forward the interests of proposed or pending legislative measures." Quoted in Hays, *Conservation and the Gospel of Efficiency,* 197–98.

55. Gifford Pinchot's continued attempts from the 1890s onward to inject wood production into national park affairs is covered in detail in Hays, *Conservation and the Gospel of Efficiency,* 189–98. He pursued this agenda with lawmakers drafting bills to establish national parks and to transfer administration of the parks to the Department of Agriculture, and he did so as part of his opposition to the establishment of the separate National Park Service. Historians have not given Pinchot's aggressive stance on this score sufficient due largely because he obscured his role by conducting private actions without public statements. As late as 1925, Henry Graves, a former chief of the agency, thought that the ultimate solution to the rivalry between the Park Service and the Forest Service would be to transfer the Park Service to the Department of Agriculture. Steen, *U.S. Forest Service,* 161.

56. Clepper notes that the agenda of the Governors' Conservation Conference in 1908 did not include wildlife, aesthetics, or watershed protection. See Clepper, *Professional Forestry in the United States,* 41–42. The role of the parks and aesthetic objectives in the Theodore Roosevelt conservation program is covered extensively in Hays, *Conservation and the Gospel of Efficiency,* esp. 122–146 and 189–98.

57. See Drummond, *Enos Mills,* and Morrison, *J. Horace McFarland.* Both Mills and McFarland were leaders of the movement for national parks.

58. Even when the Forest Service seemed reconciled to the fact that the national parks were to be administered by a separate National Park Service, its pronouncements continued to suggest that the parks would eventually be drawn on to boost the nation's timber supply. Henry Graves, Pinchot's successor as chief, wrote that "it will be a long time before the National Parks will have to be drawn upon for raw materials" but that some day the parks would serve as a supply of raw materials and until that time they should be protected. Steen, *U.S. Forest Service,* 121, 122.

59. Much of this section on the relationship between the Forest Service and Congress is drawn from the writings of Robert E. Wolf. A readily accessible example of Wolf's writings is "National Forest Timber Sales and the Legacy of Gifford Pinchot," 87–105.

60. Wolf introduces the issue by citing the challenge Pinchot's instructor in France, Professor Lucien Boppe, had given him: "When you get home to America you must manage a forest and make it pay." Quoted in Wolf, "National Forest Timber Sales and the Legacy of Gifford Pinchot," 87.

61. Clepper writes, "Pinchot noted that the economic output of the Biltmore forest was unfavorable." Clepper, *Professional Forestry in the United States,* 32.

62. Clepper, *Professional Forestry in the United States*, 32. He uses the Biltmore experience to exemplify the wider silvicultural challenge of "making forests pay."

63. See Wolf, "National Forest Timber Sales and the Legacy of Gifford Pinchot," 88, for the provision authorizing Pinchot to retain and disburse agency income. Wolf comments that there "is no hearing record on this major, last-minute, and unusual grant to spend receipts. Pinchot's subsequent testimony before the House Appropriations and Expenditures Committees makes clear, however, that he engineered this change. Congress assented because Pinchot promised to operate the reserves at a profit within five years." It was the House-Senate conference committee that added the five-year cap to the authority.

64. Wolf, "National Forest Timber Sales and the Legacy of Gifford Pinchot," 88. Wolf covers Pinchot's reports to the congressional committee in subsequent years, 1905-1910, on 88-90.

65. Wolf, "National Forest Timber Sales and the Legacy of Gifford Pinchot," 89.

66. Ibid., 90.

67. For the relationships between the agency and Congress from 1910 to 1920, see ibid., 91-92.

68. Wolf, "National Forest Timber Sales and the Legacy of Gifford Pinchot," 92-93.

69. Much of this section follows the argument in Hays, *Conservation and the Gospel of Efficiency*, which emphasizes the dominant role of decisions made by technical experts in water and land resource issues.

70. Most histories of the Forest Service focus on these dramatic incidents. See, for example, Steen, *U.S. Forest Service*.

71. It should be noted that neither of the two major pieces of legislation pertaining to the forest reserves, the authorizing act of 1891 and the management act of 1897, were products of normal legislative proceedings involving committee deliberations. That of 1891, involving a provision in a bill to repeal obsolete timber culture laws passed earlier (section 24), was inserted in conference committee, and that of 1897 was appended to the "sundry civil appropriations bill." See Clepper, *Professional Forestry in the United States*, 23, 27.

72. John Reiger describes the extensive support for beefing up enforcement of poaching regulations in Yellowstone National Park. Just a week after a major poaching incident was reported and circulated in the press, a bill introduced by Rep. John Lacey "found little opposition" and passed both houses. Reiger contrasts this action with the reluctance of Congress to move on forest reserve issues: "The reason for this," writes Reiger, "was simply that the battle over the park took place in a public arena against determined Western opposition, while the results in forestry were achieved by circumventing the popular forum." Reiger, *American Sportsmen and the Origins of Conservation*, 163-65.

73. A brief account of the Lacey Act of 1900 is in Trefethen, *American Crusade*

for Wildlife, 122ff. The standard account of the Antiquities Act is Rothman, *Preserving Different Pasts.*

74. A full-scale biography of Rep. John Lacey is not available in print as of this writing. A short biography by Greg Beisker appears on the Web site of the Iowa Natural Heritage Foundation, and there is a doctoral dissertation, "John F. Lacey: Study in Organizational Politics," by Mary Annette Gallagher (University of Arizona, 1970).

75. The AFA had long been a professional organization of leaders and promoters rather than drawing members from the public at large. Its joint activity with western irrigation promoters had come to an end when a congressional committee revealed that most of the funds for the National Irrigation Congress had come from the railroads. After 1900, promoters of the Roosevelt resource program infiltrated the organization, and by 1908, Pinchot had turned to it as the main vehicle for promoting the new "conservation" policies. However, Pinchot's influence within the AFA led to disagreements over whether or not it should develop a more narrowly focused forestry program or take up the broader "conservation" program promoted by Pinchot. Pinchot did not trust the leaders of the AFA, and over the years the organization narrowed its emphasis to forestry and moved away from the Roosevelt resource program that was shaped by the Governors' Conservation Conference of 1908.

76. The development spirit of the conservation movement is emphasized in Hays, *Conservation and the Gospel of Efficiency.*

77. Clepper, *Professional Forestry in the United States,* 41–53, describes the selective nature of the Governors' Conservation Conference of 1908, its exclusion of wildlife, aesthetics, and watershed protection, and its heavy emphasis on resource development, especially water development.

78. In their private correspondence, the "development efficiency" leaders of the Roosevelt administration belittled wildlife promotion activities. In the views of Frederick Newell, head of the Bureau of Reclamation and one of the main architects of the Roosevelt conservation program, many of the state conservation commissions "had not given full consideration to the larger questions of natural resources but confined their attention mainly to State parks, wild life and recreational features." Newell wrote to Pinchot concerning conservation that "my general impression is after the first enthusiasm died down the general attitude was not particularly favorable and there was a tendency to divert the conservation movement into narrow channels of fish and game protection." Quoted in Hays, *Conservation and the Gospel of Efficiency,* 190n71.

79. For example, in 1905, Pinchot organized the American Forest Congress as a strategy to overcome congressional opposition to the transfer of the reserves. This strategy is given only limited importance in historical writings about the transfer but is highlighted in the agency's centennial DVD, *The Greatest Good.*

80. For the role of Pinchot in the National Conservation Association, which he organized and controlled, see Hays, *Conservation and the Gospel of Efficiency,* 181–82, and passim, and Clepper, *Professional Forestry in the United States,* 139.

81. Pinchot to Brady, as quoted in Ross, *Environmental Conflict in Alaska*, 230.
82. The preoccupation of the Forest Service with wood production to the exclusion of other forest activities was reflected in a paper prepared by Chief Forester Henry S. Graves, "A Policy of Forestry for the Nation," described as "The Statement of a Policy Presented before Forestry Conferences of 1919." Recreation and watershed protection are mentioned without elaboration on two occasions in the paper, an indication of their relative unimportance in the management of the national forests. See a copy of this paper in the "Forest History" file, Hays Collection, EA.
83. Rodgers, *Bernhard Eduard Fernow*, 308.
84. Pinchot, *Breaking New Ground*, 203-12.
85. Ibid.

CHAPTER 3. Evolution of an Agency Clientele, 1920–1975

1. Quoted in Clepper, *Professional Forestry in the United States*, 199.
2. An extensive account of the Forest Service's three-decade long advocacy of public regulation of the private forest industry is in Clepper, *Professional Forestry in the United States*, chaps. 10 and 11. Chapter 10 outlines the idea of a timber famine as the prelude to regulation and chapter 11, the debate over regulation itself.
3. For the annual report by McArdle reflecting an end to the agency's "crusade" in 1952, see Clepper, *Professional Forestry in the United States*, 163.
4. For growing industry interest in replanting, see Clepper, *Professional Forestry in the United States*, 216-18.
5. Clepper emphasizes the shift toward pulp and paper as the key to making wood products investors accept replanting. See Clepper, *Professional Forestry in the United States*, 197-212.
6. Langston notes that "sales policies on the Blues [mountains of eastern Oregon] developed out of silvicultural concerns, but were simultaneously shaped and undercut by markets and technology." Langston, *Forest Dreams, Forest Nightmares*, 199.
7. Langston describes an early case in eastern Oregon when the Forest Service drew up an attractive contract that enabled the purchaser to finance a railroad to transport the logs to the mill. Langston, *Forest Dreams, Forest Nightmares*, 182.
8. See Langston, *Forest Dreams, Forest Nightmares*, 190-92, for the role of mills in determining the size and timing of timber harvests.
9. For nondeclining even flow and its role in stabilizing the forest industry, see Hirt, *Conspiracy of Optimism*, 263.
10. Clepper emphasizes the role of federal grants to states for fire and replanting in reducing controversy between the Forest Service and industry. See Clepper, *Professional Forestry in the United States*, 95-97.
11. See Clepper, *Professional Forestry in the United States*, 93-95, and Steen, *U.S. Forest Service*, 185-95, for fire provisions in the Weeks Act and the Clarke-McNary Act.

12. For the role of the CCC in forestry programs, see Clepper, *Professional Forestry in the United States*, and Steen, *U.S. Forest Service*, 213–18.

13. The standard history of the State and Private Forestry Division is Robbins, *American Forestry*.

14. Details about the Cooperative Sustained-Yield Program units are in Clepper, *Professional Forestry in the United States*, 284–85. While only one agreement between private industry and the agency was made, six were developed involving federal timber sources alone. See a brief comment on the one agreement (the Simpson unit) by forester Gordon Robinson in *Forest and the Trees*, 43–44. For unsuccessful negotiations over a proposed joint federal-private unit encompassing the communities of Libby and Troy, Montana, see Kaufmann and Kaufmann, "Toward the Stabilization and Enrichment of a Forest Community." For the problems arising from one of the units involving federal timber alone, see Forrest, "The Vallecitos Federal Sustained-Yield Unit."

15. The early attempts to establish a firmly based lumber industry in Alaska are described briefly in Ross, *Environmental Conflict in Alaska*, 230–33. A good account of the post–World War II timber contracts and the succeeding controversies over them is in Durbin, *Tongass*.

16. Hirt, *Conspiracy of Optimism*, describes many aspects of the shift from "sustained yield" to "maximum yield" in agency wood production objectives in the years after World War II. For example, the 1948 report of the chief forester says, "In handling the national-forest timber resource the Forest Service is working toward intensive management for maximum continual production." Quoted in ibid., 57.

17. For the "allowable cut effect," also called the "earned harvest effect," see Hirt, *Conspiracy of Optimism*, 264, where he describes this as a policy to "permit increases in harvest levels based on intensified management practices such as reforestation, thinning, and tree improvement."

18. The ongoing agency insistence that it could satisfy all claims in a multiple-use formula simply by increasing output for each one is the primary theme of Hirt's book, *A Conspiracy of Optimism*.

19. According to David Clary, former agency historian, "The industry kept up criticism of the Forest Service, accusing the agency of timidity in dealing with environmental interests and of not doing enough to meet industry's needs. The industry wanted an alliance with the Forest Service, but on its own terms. Too, few of the industry's leaders were willing to admit the legitimacy of competing claims, especially for wilderness." Clary, *Timber and the Forest Service*, 174.

20. Agency affirmations of the "primacy of timber" among the various forest uses and users appeared on many occasions. For example, the agency's Christopher M. Granger told a member of Congress in 1948 that "aside from relatively small areas having special value for scenic or other purposes, the national forests are managed for

timber production." Assistant Chief Edward Cliff, in a speech to the American Forestry Congress in 1953, declared, "Timber production is given priority over other uses on the most important areas of commercial forest land, with recreation, livestock grazing and wildlife being integrated as fully as possible without undue interference with the dominant use." Quoted in Clary, *Timber and the Forest Service*, 150, 151.

21. See Clepper, *Professional Forestry in the United States*, 197–286, and Devall, ed., *Clearcut*, for the growth of industrial forestry.

22. See Clepper, *Professional Forestry in the United States*, 197–286.

23. For an account of the development of higher yielding tree seeds, see Boyd and Prudham, "Manufacturing Green Gold," 107–39.

24. The decision in the Monongahela case is usually taken as the beginning of a "divorce" between the Forest Service and the environmental community. However, in a more fundamental sense it was the end of an extensive development of "industrial forestry" that had already taken place and was just becoming visible. The break between the agency and the environmental community had actually come much earlier, just after World War II, and it came at the initiative of the Forest Service when it reconsidered the boundaries of wilderness and other protected areas. It thereby gave rise to a vigorous skepticism as to the agency's reliability in supporting a continuous wilderness program.

25. The beginnings of this "industrializing" process in forest affairs was worked out by Nancy Langston from the earliest days of the Forest Service. She writes, "Much of what was wrong in the Blue Mountain forests, and in land management across the nation, came from the problems of trying to simplify and control the bewildering complexity of the natural world." Langston, *Forest Dreams, Forest Nightmares*, 156.

26. Rowley, *U.S. Forest Service Grazing and Rangelands*, is the standard work on that subject.

27. Langston, *Forest Dreams, Forest Nightmares*, 207–8, 212.

28. Rowley writes that the Forest Service was "fighting a losing battle to keep permits from assuming value in the commercial grazing community." Grazing fees tended to be higher on non–Forest Service land; at the same time, ranch land associated with national forest land in a combined livestock operation tended to be valued at a higher rate by lenders. As a result, those who held agency permits had an economic advantage over those who did not. See Rowley, *U.S. Forest Service Grazing and Rangelands*, 118–19.

29. Langston, *Forest Dreams, Forest Nightmares*, 207–10, describes the role of stockmen's associations in shaping grazing policies, especially in resisting reductions in stocking levels.

30. The Forest Service pulled back from its redistribution policy, and in 1951, Secretary of Agriculture Charles F. Brannan ordered the Forest Service to end redistribution, which was done after 1953. This change led to the gradual decline in the

number of small stockmen during the next two decades. For the redistribution policy in general, see Rowley, *U.S. Forest Service Grazing and Rangelands,* 158–59.

31. For pressures from livestock permittees to turn their permits into "property rights" subject to sale or transfer, see Rowley, *U.S. Forest Service Grazing and Rangelands,* 89. Agency officials were especially fearful that stock owners sought to make "range improvements" such as fencing and watering facilities as a move toward asserting property rights, since these improvements represented property investments.

32. Rowley comments briefly on grazing in the eastern forests. Under the Weeks Act of 1911, authorizing the purchase of eastern forest land, "grazing by domestic livestock in the East was either extremely restricted or prohibited in order to protect the other values of greater importance from a public standpoint." Rowley, *U.S. Forest Service Grazing and Rangelands,* 210. "By the early 1940s the regional forester for Region 8 concluded that sheep, swine and cattle grazing was detrimental to the production of long-leaf pine and hardwoods." Ibid., 194.

33. A comprehensive account of the growth of recreation in the national forests has yet to be written.

34. Historians have examined neither the background nor the results of the Term Permit Act of 1915. The agency program following this act was promoted for its economic benefit rather than aesthetic objectives, and the latter were pursued largely to further the economic objectives. The Forest Service had looked to the proceeds from summer homes and resorts as an important source of revenue; the developments remained in official favor because they provided a "steady source of revenue." See Steen, *U. S. Forest Service,* 153, 159–60. Just how much and thus how important this income was have not been determined.

35. Wolf, *Arthur Carhart;* Meine, *Aldo Leopold,* 156.

36. Baldwin, *Quiet Revolution,* 12–14.

37. For a chronicle of Carhart's life and work, see Wolf, *Arthur Carhart.* An account of his work with the Forest Service is Baldwin, *Quiet Revolution.* It does not, however, cover Carhart's life after his resignation from the agency but is the best detailed account of his two years of work there. Harold Steen briefly reports Carhart's frustration and resignation in *U.S. Forest Service,* 154.

38. Carhart's proposal, "Recreational Development of the Superior National Forest," has been digitized and is on the Web site of the Forest History Society. For the agency rejection of Carhart's proposals for the Quentico-Superior area in northern Minnesota, see Baldwin, *Quiet Revolution,* 100–106. Carhart's letter of resignation is reproduced in ibid., 273–78.

39. Quoted in Wolf, *Arthur Carhart,* 37, 43n10.

40. Appointment of Franklin Waugh to conduct the study and his visit to Grand Canyon with Aldo Leopold as part of his investigation are reported by Meine in *Aldo Leopold,* 159. For Waugh's contribution, see Steen, *U.S. Forest Service,* 120. For his report and comments, see Baldwin, *Quiet Revolution,* 14–16.

41. Despite the press of recreationists or "car campers" to whom the Forest Service felt compelled to respond, the agency remained somewhat skeptical and negative about them. Some agency personnel derisively referred to them as searching for a "primitive way of life." As historian Paul Hirt puts it, amid increasing demands for and conflicts over recreation, "the agency tried to contain the battles and keep the decisions from being carried over their heads, but to no avail." Hirt, *Conspiracy of Optimism*, 67.

42. A brief but useful account of the development of outdoor recreation between the wars is in Sutter, *Driven Wild*, chap. 2. Sutter cites data about the increase in the number of registered automobiles from 8,000 in 1900 to 1 million in 1913, 10 million in 1922, and 23 million in 1929. This growth was assisted by the Federal Aid Road Act of 1916, which set off a push to construct hard-surfaced highways throughout the country.

43. John Spencer, a forester for the Rocky Mountain region, quoted in Hirt, *Conspiracy of Optimism*, 65.

44. Hirt, *Conspiracy of Optimism*, 65, 67.

45. A brief overview of outdoor recreation leading up to and following the Land and Water Fund program, established by Congress in 1964, is on the Web site of the National Park Service under the title, "A Quick History of the Land and Water Conservation Fund Program."

46. The Appalachian Trail, as the "model" national trail, contains a "protected corridor" or buffer zone of 270,000 acres, most of which is a "visual" corridor.

47. At first the industry catered to hikers, backpackers, and campers. With the advent of all-terrain vehicles, a sharp division arose between those recreationists and the traditional hiking/camping group.

48. Rivalry between the National Park Service and the U.S. Forest Service in the years after World War I is the subject of Rothman, "'Regular Ding-Dong Fight.'"

49. For the struggle to create the Olympic National Park out of the Olympic National Forest, see Twight, *Organizational Values and Political Power*.

50. In 1925, the National Park Service created an education division under the direction of its chief naturalist, Ansel Hall. An analysis of the parks in 1925 by Charles Adams noted that the parks were not "devoted to technical research, but in the main to elementary educational work with the park visitors." Such work included nature walks and evening camp-fire programs and were forerunners of Park Service interpretation activities. See Sellers, *Preserving Nature in the National Parks*, 86.

51. Professional foresters long approached public criticism of the influence of wood production in forest management by complaining that the public did not understand forestry. It countered that sentiment with the need to "educate" the public. A rather remarkable exception to this persistent emphasis on educating the public to accept wood production was an analysis of a Pennsylvania attitude survey that ranked wildlife and water quality objectives high and wood production low. The authors, Steve Jones and Jim Finley of the forestry division of the extension program at Penn

State University, recommended to their fellow foresters, "As we carry out our role as professional stewards, we must consider the feelings and attitudes that the general public and landowners have toward the forest.... Perhaps it is time to stop insisting that we need to educate the public, and take time instead to educate ourselves." Steve Jones and Jim Finley, "Policy and Management Implications: The General Public and Forestry Issues," reprint of a survey report, copy in Hays, *Forest Papers*, article no. 47, 3:593–98 (copy in Hays Collection, EA).

52. See Trefethen, *American Crusade for Wildlife*; Reiger, *American Sportsmen and the Origins of Conservation*.

53. Despite its slow acceptance of wildlife responsibilities, the Forest Service from an early date authorized hunting "pack trips" into its back country. The history of such trips has not been charted, but several works recognize their importance. For example, commercial guides and packers played a prominent role in the effort to make the Lincoln-Scapegoat country of western Montana the first de facto unit of the Wilderness System, or the entire range of wilderness areas in various federal agency land holdings, established in 1964. See Roth, *Wilderness Movement and the National Forests*, 29–36. In another example, Mike McCloskey describes his efforts among hunters in southwestern Colorado to persuade them to urge their congressional representative, Wayne Aspinall (D-Colorado), chair of the House Committee on Interior and Insular Affairs, to support the wilderness proposal. See McCloskey, *In the Thick of It*, 38–41. Also, for many years the American Forestry Association sponsored pack trips into the national forests.

54. See Trefethen, *American Crusade for Wildlife*, 175, where he writes, "There were, in the early 1920's, an estimated six million licensed sport hunters, double the number in 1910." This growth significantly changed the profile of those Americans who became involved in hunting. Trefethen continues, "By the time of the Harding Administration, thousands of hunters from the laboring and middle classes were competing with the wealthy sportsmen in the hunting fields."

55. Trefethen, *American Crusade for Wildlife*, 217–29, describes the growth of the organizational infrastructure for sport hunting that developed in the 1930s.

56. For the role of the Izaak Walton League in the campaign to protect the recreational area of the Superior National Forest, see Baldwin, *Quiet Revolution*, 147–51.

57. For state game management, with a central focus on the enforcement of game laws, see Trefethen, *American Crusade for Wildlife*, 237–42.

58. Leopold's wildlife activities in the Southwest are described in Meine, *Aldo Leopold*, 135–54.

59. Meine, *Aldo Leopold*, 135.

60. Quoted in ibid.

61. Hirt, *Conspiracy of Optimism*, 60–63.

62. Rowley, *U.S. Forest Service Grazing and Rangelands*, 77–78.

63. For the resistance of the livestock committees to reductions in stocking levels, see Langston, *Forest Dreams, Forest Nightmares*, 207-10.

64. This controversy between the Forest Service and the state fish and wildlife agencies is chronicled in Catton and Mighetto, *Fish and Wildlife Job on the National Forests*, 79-109.

65. Trefethen, *American Crusade for Wildlife*, 217-29.

66. Hirt, *Conspiracy of Optimism*, 69-63.

67. The response to the 1939 report by foresters is dealt with briefly in Wilkinson and Anderson, *Land and Resource Planning in the National Forests*, 283-84; quotation is from 283n1505.

68. Swift's oral history, "Wildlife Policy and Administration in the U.S. Forest Service," conducted in 1964 by Amelia Fry, Regional Oral History Office, Bancroft Library, University of California, revised 1975, 10-11, as quoted in Hirt, *Conspiracy of Optimism*, 63. Swift was concerned that the agency consigned wildlife management to a marginal position within its organization.

69. The transition from game to nongame to biodiversity is traced in Hays, *Wars in the Woods*, chap. 2. One event that marked the early stages of this transition was when the North American Game Policy, drawn up by wildlife (game) leaders, was revised in 1963 and the title changed to North American Wildlife Policy.

70. For Leopold's interest in watersheds, see Meine, *Aldo Leopold*, 188-89, 210, 220.

71. For Leopold's analysis of long-term changes in agricultural river valleys in Arizona, see Meine, *Aldo Leopold*, 188-89.

72. For Leopold's focus on riparian areas, in contrast with headwater areas, see Meine, *Aldo Leopold*, 188-89.

73. See Meine, *Aldo Leopold*, 204-10, for Leopold's observations on the Prescott National Forest.

74. Quotations from Meine, *Aldo Leopold*, 204.

75. Quoted in ibid., 210.

76. Meine, *Aldo Leopold*, 220.

77. Quoted in Meine, *Aldo Leopold*, 217.

78. See the agreement between Charles Cherrington, mayor of Grand Junction, Colorado, and the Forest Service, June 16, 1915 (copy in author's files). In it, the mayor and Forest Service approved the use of the relevant acres of the adjacent national forest, except for "measures necessary for the proper protection and care of the forest; the marking, cutting and disposition of such timber as in the judgment of the Forest officers may be removed without injury to the water supply of said city."

79. Hirt, *Conspiracy of Optimism*, 38.

80. Ibid.

81. Quoted in Clary, *Timber and the Forest Service*, 164.

82. Hirt, *Conspiracy of Optimism*, 224.

83. For the emerging roles of water quality, the salmon fishery, and acid precipitation in forest management, see Hays, *Wars in the Woods*, 92–93, 147–48. In the western forests, both state and national, the issue was more intense in the three Pacific coast states of the lower forty-eight.

84. Harold Steen describes the debate within agency leadership as to the relative wisdom of cutting or not cutting timber in travel corridors. He describes a staff meeting on the subject in 1924. Herbert Smith, responsible for public information, recounted the opposition to Fernow in New York when he attempted to foster logging in areas thought of as scenic by the public. At the meeting some denounced "extremists" for demanding "scenic strips" along all roads. Another argued that "it did not matter what the Forest Service believed; most of the public favored recreation," concluding, "we should be flying against present public sentiment if we adopt a policy of cutting straight to the road." Steen, *U.S. Forest Service*, 158–59.

85. The rivalry between the National Park Service and the Forest Service is a theme of many accounts. A useful one is Rothman, "'Regular Ding-Dong Fight.'"

86. A history of the Great Smoky Mountains National Park in a context larger than the park itself is Margaret Lynn Brown, *The Wild East: A Biography of the Smoky Mountains* (Gainesville: University Press of Florida, 2000). For Shenandoah, see Darwin Lambert, *The Undying Past of Shenandoah National Park* (Boulder, CO: Roberts Rinehart, in cooperation with Shenandoah Natural History Association, 1989).

87. Twight, *Organizational Values and Political Power*.

88. Hirt, *Conspiracy of Optimism*, 226.

89. The Forest Service wilderness program prompted more writing than any other agency activity. The classic account of the evolution of wilderness as an idea is Nash, *Wilderness and the American Mind*. A brief but succinct account of the politics of wilderness, especially its administrative twists and turns, is Scott, *Enduring Wilderness*. Especially useful is his treatment of "de facto" wilderness, including the eastern wilderness areas, and the history of wilderness protection on land managed by other agencies. See also Gilligan, "Development of Policy and Administration of Forest Service Primitive and Wilderness Areas in the Western United States."

90. Biographical accounts of these three wilderness figures are Meine, *Aldo Leopold;* Glover, *Wilderness Original;* Baldwin, *Quiet Revolution*.

91. Schrepfer, "Establishing Administrative 'Standing,'" 127.

92. The Forest Service has developed its own interpretation of its role in wilderness history. It gives considerable agency credit to the origin of the wilderness program by featuring its two employees, Aldo Leopold and Bob Marshall, as pioneers. See, for example, Roth, *Wilderness Movement and the National Forests* (Roth was at one time the official historian of the Forest Service), and the centennial DVD, *The Greatest Good*. These interpretations are very selective and say little about the road built through the first wilderness, the Gila, in 1929, the degree to which the agency made

clear that wilderness was acceptable only if it did not impede wood harvest, the roadblocks which it continually placed in the way of advances in wilderness, and the break with wilderness advocates when, after World War II, it took steps to shrink the boundaries of areas it had previously identified as protected.

93. The "U" regulations refined the "L" regulation; both identify different types of protected forest areas.

94. Leopold's specific proposal for wilderness on the Gila National Forest arose from his work with sportsmen in New Mexico on a variety of game management issues. See Meine, *Aldo Leopold*, 145–46, 154. According to Leopold, "By wilderness I mean a continuous stretch of country preserved in its natural state, open to lawful hunting and fishing, big enough to absorb a two-weeks' pack trip and devoid of roads, artificial trails, cottages, or other works of man." Meine, *Aldo Leopold*, 196, quoting from Leopold's article, "The Wilderness and Its Place in Forest Recreation Policy," *Journal of Forestry*, November 1921. See also Baldwin, *Quiet Revolution*.

95. Leopold later described the incident and partly blamed himself for it because of his cooperation in eradicating predators. "Here my sin against the wolves caught up with me. The Forest Service, in the name of range conservation, ordered the construction of a new road splitting my wilderness in two, so that hunters might have access to the top-heavy herd. I was helpless . . . I was hoist on my own petard." Quoted in Huggard, "America's First Wilderness Area," 154.

96. Quoted in Huggard, "America's First Wilderness Area," 143.

97. For the Superior National Forest wilderness issue, see Baldwin, *Quiet Revolution;* Proescholdt, Rapson, and Heinselman, *Troubled Waters*.

98. In a report to the National Conference on Outdoor Recreation by a joint committee of the American Forestry Association and the National Parks Association in 1924, the twenty-one proposed wilderness areas comprising 12.5 million acres were described as the "least accessible and least commercially valuable tracts." Quoted in Steen, *U.S. Forest Service,* 153. An account of the political maneuvering over one wilderness issue, the French Pete in Oregon, is Andy Kerr, "The Browning of Bob Packwood," posted on Andy Kerr's Web site.

99. Historian Susan Schrepfer covers the change in the Forest Service's approach to wilderness in *Nature's Altars,* chap. 8.

100. Quoted in Scott, *Enduring Wilderness,* 38.

101. For a brief account of the debate over the Wilderness Act, see Scott, *Enduring Wilderness,* 37–56.

102. For de facto wilderness areas, see Scott, *Enduring Wilderness,* 62–66, 79–83.

103. Scott, *Enduring Wilderness,* 68–72. As of yet, there is no comprehensive historical account of eastern wilderness areas.

104. For the Forest Service "virgin wilderness" argument, see Scott, *Enduring Wilderness,* 66–68.

105. The manual has been updated several times. See John C. Hendee and Chad

P. Dawson, *Wilderness Management: Stewardship and Protection of Resources and Values*, 3rd ed. (Golden, CO: WILD Foundation and Fulcrum Publishing, 2002).

106. The results of the survey's first two years are discussed in Brett J. Butler and Earl C. Leatherberry, "America's Family Forest Owners," *Journal of Forestry* 102, no. 7 (Oct-.Nov. 2004): 4–9.

107. See Hirt, *Conspiracy of Optimism*, 171–92. Hirt's chapter, "Multiple Use and Sustained Yield: Debated and Defined, 1955–1960," can be paraphrased succinctly: the Forest Service interest was largely to keep decisions about the balance among uses in their own hands.

108. In contrast with many historians who find the roots of the interest in forests in "prominent" writings, I seek those roots in people's personal experience with forests and therefore place considerable emphasis on the increasing numbers of people who took part in activities in forests or who had "vicarious" contact with forests through reading and visual imagery.

109. The significance of planning for each national forest, as required by the National Forest Management Act of 1976, is outlined in Hays, *Wars in the Woods*, 35–40.

110. While outdoor recreation is referred to frequently in books about the national forests, its complex growth over the years and the relationship between its promoters and its participants has not yet been adequately explored.

111. On the Wilderness Preservation Act's designation of areas, see Scott, *Enduring Wilderness*. A far more detailed account is in Roth, *Wilderness Movement and the National Forests*.

112. The first citizen-proposed de facto wilderness is described briefly in Scott, *Enduring Wilderness*, 79, and more extensively in Roth, *Wilderness Movement and the National Forests*, 29–36. A more popular telling of the Lincoln-Scapegoat story is Tom Price, "From Hardware to Software: How the Wilderness Movement Got Its Start," *High Country News* (Paonia, Colorado), 33, no. 2 (January 29, 2001): 20.

113. After 1960 there was tremendous growth in nongame wildlife programs at the state level. Nongame wildlife uses were also known as "nonconsumptive wildlife uses" and often were defined as "wildlife not legally hunted." The nongame wildlife programs may be viewed not only in their own right but as a bridge between the earlier interest in hunted wildlife and the later interest in biodiversity.

114. In a few cases that involved the geographic extent of species habitat, the Forest Service established zoning, primarily in the form of protecting birthing areas (e.g., for deer and black bears) or bird nesting areas (e.g., for eagles and spotted owls); the zoning practice did not extend much further. When the use of the radio collar greatly expanded knowledge about the range of movement of large mammals and birds, and when the interest in biodiversity brought animals of very limited movement into the wildlife scheme of things, zoning seemed to be far less feasible, hence the controversy over the "survey and manage" philosophy in the Pacific Northwest, which was one ap-

proach to dealing with these relatively immobile animals. "Gap analysis," pioneered by botanist Michael Scott, envisaged a more expanded version of zoning to identify areas containing large numbers of common species for protection under the objective of "keeping common species common." Although many public land agencies joined in the general approach of gap analysis, few actually designated land areas to be subject to such zoning.

115. A great number of books have been written about the salmon issue in the Pacific Northwest. Among them are David R. Montgomery, *King of Fish: The Thousand Year Run of Salmon* (Boulder, CO: Westview Press, 2003); Jim Lichatowich, *Salmon without Rivers: A History of the Pacific Salmon Crisis* (Washington, DC: Island Press, 1999); and Molly Cone, *Come Back Salmon: How a Group of Dedicated Kids Adopted Pigeon Creek and Brought It Back to Life* (San Francisco: Sierra Club Books for Children, 1992).

116. The Forest Service's optimism that it could satisfy everyone by increasing the "benefits" of national forest management for all is the theme of Hirt's *Conspiracy of Optimism*.

117. Some saw timber production as the way to approach all other uses. The timber management plan for the Willamette National Forest, prepared in 1965, declared that the timber manager "should direct his management towards the use of timber harvesting as a tool in attaining true multiple-use management." Clary, *Timber and the Forest Service*, 172.

118. Quoted in Clary, *Timber and the Forest Service*, 151.

119. Planning in the 1980s, including the use of FORPLAN, and the role of Randall O'Toole's critiques and his publication, *Forest Planning* (copies of this newsletter in Hays Collection, EA), are described in Hays, "New Environmental Forest"; see esp. sec. III, "New Values and Objectives," and sec. IV, "Incorporating Environmental Objectives into Planning." See also Hirt, *Conspiracy of Optimism*, 274–75.

120. The role of John Crowell in seeking to enhance wood production in the national forests during the Reagan administration is described in O'Toole's publication, *Forest Planning*. At a convention in 1989, the supervisors composed a letter to the agency chief complaining that the "timber program has been 35 percent of the National Forest System (NFS) budget for the last 20 years while recreation, fish and wildlife, and soil and water have been 2 to 3 percent each." Quoted in Hirt, *Conspiracy of Optimism*, xli.

121. Details and implications of the evolution of "sustained yield" into "maximum yield" are described briefly in Hirt, *Conspiracy of Optimism*, 132–34.

122. A significant resource for this section is Wolf, "National Forest Timber Sales and the Legacy of Gifford Pinchot," 93–95.

123. Wolf, "National Forest Timber Sales and the Legacy of Gifford Pinchot," 93.

124. A brief account of the agency trust funds can be obtained on the Web site of

Taxpayers for Common Sense (http://www.taxpayer.net), under its project titled "Forest Service Budget Reform," where the history of each fund is described separately. When Forest Service chief Michael Dombeck resigned in 2001, in his letter of resignation he recommended that the trust funds be abolished so that all of the agency's divisions, for budget purposes, should be placed on an "even keel."

125. For the GAO report see Wolf, "National Forest Timber Sales and the Legacy of Gifford Pinchot," 97.

126. Hirt, *Conspiracy of Optimism*, 188.

127. Forest Service attempts to obtain support from the Sierra Club are described in Schrepfer, "Establishing Administrative 'Standing.'"

128. See the bonus features (sections 2 and 3) of *The Greatest Good* DVD for pictures of a great number of Smokey Bear advertisements, many of them produced by the Advertising Council of America.

129. See Steve Jones and Jim Finley, "Policy and Management Implications: The General Public and Forestry Issues," copy in Hays, *Forest Papers*, article no. 47, 3:589–94.

130. Quoted in Meine, *Aldo Leopold*, 126.

131. Meine, *Aldo Leopold*, 126.

132. A detailed study of the impact of these changes on grazing in the Coronado National Forest in the Southwest is Diana Hadley, "Grazing the Southwest Borderlands."

CHAPTER 4. Confronting the Ecological Forest, 1976–2005

1. This section draws heavily from Hays, *Wars in the Woods*, chap. 2, which emphasizes the impact of ecological objectives on national forest management. Understanding these changes requires a heavy focus on individual national forests rather than Washington, DC–based national policy.

2. The extensive and unique biological resources of the national forests were emphasized by some agency staff in a publication of the botany section of the Watershed, Fish, and Wildlife Division (WFW) of the Forest Service, titled *Lingua Botanica*. The editor declared in its first issue, "These pages are the fruit of efforts put forth by a group of people that are passionately dedicated to the grand botanical heritage of this nation's public lands." See *Lingua Botanica*, which was first published online in 2000 and was discontinued in 2005.

3. An attempt to understand the perceptions and values underlying this widespread interest in nature on a fundamental level is Kempton, Boster, and Hartley, *Environmental Values in American Culture*. The authors identify the "mental maps" through which people understand the world around them. See esp. chap. 3 in that volume, "Cultural Models of Nature."

4. A wide range of publications and activities reflect this discovery of and interest in the flora and fauna of the nation's wild lands. See, as examples, Maine Natural Resources Council, "Maine's Natural Heritage... Life Depends On It! Protecting Maine's Plant and Animal Legacy," a statement supporting the establishment of an ecological reserve system for Maine (Augusta, Maine, January 2000); *Chicago Wilderness: Exploring Nature and Culture*, published quarterly beginning in 1996 by Chicago Wilderness, "an unprecedented alliance of 182 public and private organizations working together to study and restore, protect and manage the precious natural ecosystems of the Chicago region for the benefit of the public." For a focus on biodiversity, especially in the national forests, see "Getting on Message: Eastern Forests and Biodiversity," released by the Biodiversity Project (Madison, WI), in 2002. For a study of the attitudes of Americans toward biodiversity, see *Americans and Biodiversity*, a survey conducted by Belden Russonello & Stewart for the Biodiversity Project (Washington, DC, April 2002).

5. For a more extensive treatment of this "ecological discovery" in the national forests see Hays, *Wars in the Woods*, chap. 2.

6. The clash of these two forest "visions," one environmental/ecological and the other commodity based, is the theme of Hays, *Wars in the Woods*.

7. See Opinion Research Corporation, *The Public's Participation in Outdoor Activities and Attitudes toward Wilderness Areas* (Princeton, NJ, 1977). For example, to the question, "Do you think the U.S. Forest Service should try to increase the yield and sale of timber from the national forests or should it continue to preserve these trees in their natural state?" 28 percent agreed with the "increase yield option" and 62 percent with the "preserve these trees" option.

8. For an example of a state survey, see Jones and Finley, "Policy and Management Implications"; Myron R. Schwartz, "The Follow-Up Survey for the Analysis of Attitudes and Knowledge of the Stewardship Program in Pennsylvania: Summary Statistics" (copies of both in Hays, *Forest Papers*).

9. For the southern Appalachian survey conducted by the Forest Service, see *Public Survey Report: Southern Appalachian Forest; A Survey of Residents of the Greater Southern Appalachian Region to Describe Public Use and Preferred Objectives for Southern Appalachian National Forest* (Southern Region of National Forest Systems, FS; Southern Research Stations, FS; University of Tennessee, July 2002). For the Forest Service survey of national attitudes on forest issues, see Deborah J. Shields, Ingred M. Martin, Wade E. Martin, and Michelle A. Haefele, *Survey Results of the American Public's Values, Objectives, Beliefs, and Attitudes Regarding Forests and Grasslands: A Technical Document Supporting the 2000 USDA Forest Service RPA Assessment*, Gen. Tech. Rep. RMRS-GTR-95, Fort Collins, CO, U.S. Department of Agriculture, Forest Service, Rocky Mountain Research Station.

10. Professional organizations with an ecological forestry focus are described briefly in Hays, *Wars in the Woods*, 91.

11. For an example of an ecological perspective on eastern forests, see Yahner, *Eastern Deciduous Forest.*

12. The work at the H. J. Andrews Experimental Forest (or Andrews Forest) located in the Willamette National Forest in Oregon is covered extensively in Luoma, *Hidden Forest.* The work in the Andrews Forest can be followed on its Web site. Professor Jerry Franklin of the University of Washington was the most well-known public figure in advancing the ecological forestry ideas generated from research at the Andrews Forest.

13. See the curriculum program entitled "The Secret Forest Experience Project" (Forest Service Employees for Environmental Ethics, Eugene, OR, 2001). This curriculum was designed for students and has special emphasis on biodiversity, terrestrial forest arthropods, fire ecology, and forest soil microorganisms

14. Planning and management for several forests in a regional context (e.g., the Greater Yellowstone area, the Sierra Nevada, and the Pacific Northwest) have brought together both citizen groups and scientists.

15. *The Greatest Good* DVD, issued in 2005, did not acknowledge either the public's or scientific community's interest in the biodiversity content of the national forests.

16. Citizen groups interested in ecological forestry objectives, most of which were organized after the passage of the National Forest Management Act of 1976, along with their many ways of formulating and promoting their objectives, are described in Hays, *Wars in the Woods,* chap. 2.

17. The ecological interests of scientists are described in Hays, *Wars in the Woods,* chap. 2.

18. These were frequently called "survey and manage" procedures, namely that prior to harvest, areas should be surveyed so as to establish an initial condition from which potential changes due to harvest could be measured.

19. For the Massabesic Survey, see Northeastern Research Station, "Vegetation of Forested Uplands in the Massabesic Experimental Forest," Gen. Tech. Rep. NE-320. This is a summary of an inventory of five hundred species and subspecies of vascular plants.

20. The increasing use of radio telemetry to identify habitat for wide-ranging animals and birds, first for grizzly bears in Yellowstone National Park and then for raptors, whooping cranes, and butterflies, marks a major development in wildlife knowledge and management. While its use in specific cases is the subject of many scientific articles, such as those pertaining to the spotted owl, it has yet to receive attention as a major phase in the history of wildlife management. For a review of the different kinds of radio telemetry and its application to different kinds of animals and birds, see Jan Welsh and Bob Welsh, "Wired Life" (Minnesota Department of Natural Resources nongame division, Nov.–Dec. 2003). For a specific case, see "New Satellite

Collar Aids Division of Wildlife Researchers in Monitoring of Lynx in Colorado," Colorado Division of Wildlife, news release, June 29, 2000. See also U.S. Geological Service and the Wildlife Society, "Forum on Wildlife Telemetry," Sept. 21–23, 1997. For a brief account of the history of radio telemetry on animals, now called "bio-logging," see Yan Ropert-Coudert and Rory P. Wilson, "Trends and Perspectives in Animal-attached Remote Sensing," *Frontiers in Ecology and the Environment* 3, no. 8 (Oct. 2005): 437–44.

21. The entry of the EPA and the Fish and Wildlife Service into the debate over the role of timber harvests on water quality marked a new phase of the agency's traditional indifference to the watershed mandate of the Forest Management Act of 1897. The resulting disputes involved considerable state-level controversy and litigation in which both scientists and citizens challenged harvest practices in California, Oregon, and Washington. A specific point of dispute arising from these circumstances was whether or not forest activities required a water quality permit. Both private and public wood production agencies argued that only identifiable discharge sources (point sources) did so and that the "discrete conveyance points," which included ditches or culverts in wood harvest operations, were not "point sources." In one case carried to litigation in California, a federal court judge ruled that ditches and culverts were "point sources" and thus required a permit. The Forest Service was not inclined to implement this decision in its management practices. See the Web site of the Environmental Protection Information Center of Garberville, Calif., which contains the article entitled "EPIC Water Case Moves Ahead: Federal Court Reins in Logging Pollution," Dec. 1, 2003, California file, Hays Collection, EA.

22. For an overall account of non-point source water pollution and its role in "total maximum daily loads" of stream pollution, see Houck, *Clean Water Act TMDL Program*, esp. 61–63, which describes the Forest Service's objection to the EPA non-point source program. For litigation that described the impact of clear-cutting on salmon streams and the landslides it caused, see ibid., n.160.

23. Ellis B. Cowling, "Acid Precipitation in Historical Perspective," *Environmental Science and Technology* 16 (1982): 110A–23A; Ernest J. Yanarella and Randal H. Ihara, eds., *The Acid Rain Debate: Scientific, Economic, and Political Dimensions* (Boulder, CO: Westview Press, 1985).

24. For an account of the gradual understanding of the hydrological role of acid deposition, see Jenkins et al., *Acid Rain in the Adirondacks*.

25. For a succinct statement as to the role of headwaters in watershed management, see Winsor H. Lowe and Gene E. Likens, "Moving Headwater Streams to the Head of the Class," *BioScience*, March 2003, 196–97.

26. The Northwest plan, fostered by the Clinton administration, involved a major emphasis on watershed protection to benefit salmon and other fish. One of its most important elements was an interagency watershed monitoring program that was the

most comprehensive such program to date. See Gordon H. Reeves et al, "Aquatic/ Riparian Effectiveness Monitoring Plan" (AREMP), http://www.reo.gov/monitoring/ watershed/arcmp/. The interagency format of the Northwest plan soon came to an end through the failure of each agency to provide funding; the fate of the monitoring plan is not known.

27. The forest plan for the Northwest required that each individual harvest on a watershed be evaluated in terms of its impact on the entire watershed. The George W. Bush administration, on the other hand, eliminated this requirement in its revision of the Forest Management Plan on the grounds that impairment of a smaller watershed was overcome by the more comprehensive influence of the relatively unimpaired larger watershed on which no harvest had taken place and thus the impact of the individual harvest could be ignored.

28. A study that emphasized the limited significance of the best management practice (BMP) programs in many states is Steve Kahl, "A Review of the Effects of Forest Practice on Water Quality in Maine," Report to the Maine Department of Environmental Protection, Bureau of Land and Water Quality Control, Augusta, Maine, Oct. 1996, in Box 32, State Forest Affairs, Forestry List EA, Hays Collection, EA. This study reviews BMP programs in many states.

29. The scientific analysis of the Sierra Nevada situation outlined four major disturbances to the region: the impaired aquatic/riparian systems; the degree to which timber harvest had increased the severity of fires; the adverse impact of loss of riparian and forest habitat on wildlife; and the damage done to riparian areas by mining, dams, diversions, roads, logging, residential development, and recreational activities. A statement of ecological restoration forest objectives is Dominick A. DellaSalla et al., "A Citizen's Call for Ecological Forest Restoration: Forest Restoration Principles and Criteria," *Ecological Restoration* 21, no. 1 (Mar. 2003): 14–23.

30. For the Bitterroot and Biscuit fires, see Hays, *Wars in the Woods*, 261, 262, 276–77, and esp. chap. 6, notes 29, 49.

31. The West Virginia Highlands Conservancy, a citizen group, succinctly expressed its view of ecologically based forest health objectives: "For the Forest Service and the timber industry the focus in 'forest health' is on 'tree health' and especially trees that have commercial value as timber. However, a more ecologically sound definition of forest health implies much more than just trees. A forest includes many interacting plants, animals, insects, and micro-organisms that live and reproduce there. For example, the soil alone contains thousands of types of fungi, bacteria, earthworms, and insects, all of which are essential to growing future generations of healthy trees (as well as other organisms)." *Highlands Voice* (Charlestown, WV), April 2004, 18.

32. This change in agency thinking is one of the major themes of the Forest Service's centennial DVD, *The Greatest Good*.

33. A good review of the National Environmental Policy Act and its implications for environmental reviews is Lindstrom and Smith, *National Environmental Policy Act*.

34. A useful overview of the Endangered Species Act and its subsequent history is Burgess, *Fate of the Wild*.

35. For the National Forest Management Act of 1976, see LeMaster, *Decade of Change;* and Wilkinson and Anderson, *Land and Resource Planning in the National Forests,* 144–146.

36. Revision of the forest regulations adopted by the George W. Bush administration dropped the NEPA analysis as part of the forest planning process. Henceforth, the forest supervisor was to undertake the scientific analysis of forest-wide issues, which reduced markedly the opportunities for both citizens and scientists to influence the planning process.

37. In *The Greatest Good* DVD presentation, forest ecologist Jerry Franklin is shown making the point that NEPA played a crucial role in bringing a more varied group of scientists into the forest planning process.

38. The Forest Service had an ongoing concern about the large number of comments received in response to individual forest plans or larger issues such as the Clinton administration's proposal to protect roadless areas. It was particularly alarmed by the larger number of people whose replies, usually by postcards or Internet reply forms, were simply "yes" or "no" without further comment, and over the years the agency decided that only comments with written explanations were to be counted. In order to fend off criticism, it established a content analysis team to which it sent all the results of public comments for analysis. During the George W. Bush administration, however, it took a more aggressive stand by giving the team instructions as to how it should decide what comments to accept or not and then to outsource the analysis of comments to a private group outside the agency. See Greg Hanscom, "Outsourced: As the Bush Administration Rushes to Put the Public Lands into the Hands of Private Industry, a Model Group of Forest Service Employees Gets Canned," *High Country News,* Apr. 26, 2004, 7–12.

39. Nancy Manring, "Democratic Accountability in National Forest Planning," *Journal of Forestry* 102 (Mar. 2004): 43–47, pertaining to citizen-agency relationships, in a special issue titled "Appeals, Litigation and Forest Policy."

40. The new forest regulations promulgated by the George W. Bush administration sought to increase the decision-making autonomy of forest supervisors by reducing the role of citizen groups in the formulation of forest plans, freeing them from challenges by other government agencies such as the EPA and the U.S. Fish and Wildlife Service and eliminating the requirement for the use of peer groups in evaluating scientific issues. See "Forest Regulation Revision" file in Hays, Forest List 3, Hays Collection, EA.

41. This role of the WFW in providing assistance for "compliance requirements" rather than management objectives was described on its Web site by its new director, Dr. Deanna J. Strouder, in May 26, 2005.

42. The role of the appeals process in the Forest Service during the twentieth

century, debate over it in the last few decades of the century, and the change made in the Healthy Forests Act of 2004 are the subject of Vaughn and Cortner, *George W. Bush's Healthy Forests*.

43. The analysis of Forest Service budget events that follows is drawn from Wolf, "National Forest Timber Sales and the Legacy of Gifford Pinchot," 96-100.

44. Forest Service budget reports during the 1970s and 1980s continued to stress total income from timber sales and to provide only meager data about cost. See Wolf, "National Forest Timber Sales and the Legacy of Gifford Pinchot," 95-96.

45. For events in the "below cost" controversy see Wolf, "National Forest Timber Sales and the Legacy of Gifford Pinchot," 96-100.

46. Thomas Barlow, Gloria Helfand, Blair Orr, and Thomas Stoel, *Giving Away the National Forests: An Analysis of U.S. Forest Service Timber Sales below Cost* (Washington, DC: Natural Resources Defense Council, June 1980).

47. V. Alric Sample, *Below Cost Timber Sales on the National Forests* (Washington, DC: The Wilderness Society, 1984). Sample's analysis pertained to the years 1978, 1982, and 1983 and provided data for each of the nine regional districts of the Forest Service. It concluded that five of the nine regions had net losses, for a total loss of $443 million for the combined three years.

48. The General Accounting Office report on timber sales is cited in Wolf, "National Forest Timber Sales and the Legacy of Gifford Pinchot," 104n43.

49. O'Toole's economic analyses of national forest finances are in the Hays Collection, EA. The major published work on his economic analysis is O'Toole, *Reforming the Forest Service*.

50. The agency accounting system developed under congressional mandate was called the Timber Sale Program Information Reporting System (TSPIRS). Its content and argument concerning its adequacy are recounted in Wolf, "National Forest Timber Sales and the Legacy of Gifford Pinchot," 97-98.

51. O'Toole quoted in Wolf, "National Forest Timber Sales and the Legacy of Gifford Pinchot," 98.

52. Wolf, "National Forest Timber Sales and the Legacy of Gifford Pinchot," 100; *Report of the Below-Cost Timber Sales Task Force, Fiscal and Social Responsibility in National Forest Management* (Bethesda, MD: Society of American Foresters, 1986), 100.

53. This "write-off" extended well beyond useful life or replacement. See the Wilderness Society, *1988 Timber Receipts and Expenditures on the National Forests, by Forest Service Region* (Washington, DC: The Wilderness Society, 1989).

54. Wolf summarizes the agency's budget situation as follows: "Gifford Pinchot's unmet challenge—to manage the national forests and make them pay—still haunts the agency." See Wolf, "National Forest Timber Sales and the Legacy of Gifford Pinchot," 100.

55. The controversies between older commodity and newer environmental ob-

jectives in national forest management are the subject of Hays, *Wars in the Woods,* chap. 6. The analysis in this section is developed more fully there.

56. The "survey and manage" issue was most extensively debated in the implementation of the Pacific Northwest Forest Plan. For an account of the issue during the administration of George W. Bush, see Hays, *Wars in the Woods,* 255-60.

57. When the Supreme Court upheld the authority of the Environmental Protection Agency to implement the non-point source provisions of its water quality program, the Forest Service sought to blunt its action by arguing that the Clean Water Act applied only to point sources. The agency agreed with the position of the states that non-point source regulation should not establish specific limits on pollution. See Houck, *Clean Water Act TMDL Program,* 61-62.

58. Considerable controversy ensued over the limited approach of the Forest Service to taking responsibility for inventorying and monitoring species in the national forests. Litigation in which citizens and scientists pressed the agency to move forward on this score cropped up in many parts of the country. The southern Appalachians was one region that generated considerable action involving the controversy over the use of management indicator species as a method of inventory and monitoring.

59. The administration's announcement of changes in its Aquatic Conservation Strategy for the Pacific Northwest stated that the nine ACS objectives "would be attained at the fifth-field watershed scale and not at the project of site level." From the Web site "Aquatic Conservation Strategy," see "Final Supplemental Environmental Statement," under the section "The Proposed Amendment," 9.

60. For the opposition of the Republican Party to the nation's environmental/conservation programs during the first term of the George W. Bush administration (2001-2005), see Robert F. Kennedy Jr., *Crimes against Nature: How George W. Bush and His Corporate Pals Are Plundering the Country and Hijacking Our Democracy* (New York: Harper/Collins, 2004). For the activities of a Republican group favorable to environmental objectives, see the Web site of Republicans for Environmental Protection, http://www.rep.org, and especially its publication, *The Green Elephant.*

61. For Dombeck's "National Resource Agenda," see Dombeck, Wood, and Williams, *From Conquest to Conservation.*

62. Bosworth's views were contained in speeches he made, especially throughout the West. For an account of one speech he gave in Denver, see the editorial "Forests Face Threats," *Denver Post,* Sept. 28, 2003; for another, see Michael Milstein, "Forest Service Chief Suggests Way to Cut through Logging Appeals," *Oregonian,* Dec. 18, 2001. A public statement of Bosworth's objectives is, "Statement, Dale Bosworth, Chief, USDA Forest Service, Subcommittee on Forest Health, Committee on Resources, U.S. House of Representatives, Washington, D.C., Dec. 4, 2001." Copies of these documents are in the "Statements by Forest Service Officials" folder, section titled "Forests Nationwide," in Hays, Forest List 2, Hays Collection, EA. A perceptive

recognition of the differences in agenda between Dombeck and Bosworth is contained in the form of editorials in the *Idaho Falls Post Register* after a visit by Bosworth with the newspaper's editorial board. See J. Robb Brady, "Bosworth's Agenda," and "Gaps in Bosworth's Goals," *Idaho Falls Post Register,* Jan. 25 and Jan. 26, 2004.

Epilogue

1. Material concerning this issue, including the relevant documents, can be found on the Internet accessed under the title "Sierra Club Sequoia Task Force."

2. Dombeck described the road situation and his approach to it in Dombeck, Wood, and Williams, *From Conquest to Conservation,* 102–16.

3. For the Forest Service Employees for Environmental Ethics, see its Web site (FSEEE) and associated archive; see also its quarterly magazine, *Forest Magazine* (1998–). For an account of one employee the group defended, see "Tongass Wildlife Biologist and Whistleblower Glen Ith Dies," *Forest Magazine* 10, no. 3 (summer 2008): 29–30.

4. The initial (but unsuccessful) call for professionals with newer specializations came in the 1920s, when landscape architects were sought. See Wolf, *Arthur Carhart,* 97–98.

5. The Web site for WFW can be reached via U.S. Forest Service–WFW. The WFW monthly newsletter has been archived since the division's reorganization in 2005.

6. For the Greater Yellowstone Clean Air Partnership, see the Web site "U.S. Forest Service, Regional Air Quality." A most useful description of the scientific and technical aspects of the program is in the job description for the air program manager for the Bridger-Teton and Shoshone National Forest under the Internet key words, "Natural Resource Specialist (Air Quality) Bridger-Teton and Shoshone National Forest."

7. See the Web sites of Forestry for the Bay and "Chesapeake Bay Program: Forest Buffer Restoration." For the forest acquisition program, see Lora Lutz, "Regional Goal Will Preserve 695,000 Acres of Forests" in *Chesapeake Bay Journal,* January 2000, archived under the journal's Web site.

8. Contrasting commodity and ecological forestry issues are outlined more fully in Hays, *Wars in the Woods,* 20–87.

9. For "survey and manage" issues, especially in the Pacific Northwest, see ibid., 164–65.

10. Considerable detail about the warbler recovery is available on the Internet under the key words "Kirtland's warbler." There is even an annual festival coinciding with the species count, which takes place when the birds return from the Caribbean in the spring.

11. A review of Bush administration attempts to reinterpret forest-related statutes

in an anti-environmental direction is in a report of the Judicial Accountability Project: William Snape III, Michael T. Leahy, and John H. Carter II, *Weakening the National Environmental Policy Act: Undercutting National Forest Protection* (Washington, DC: Defenders of Wildlife, summer 2003).

12. For the Bush attempt to undermine the role of the Fish and Wildlife Service consultation under the ESA, see "Bush to Relax Protected Species Rules," *Daily Camera* (Boulder, CO), Aug. 13, 3008, 5A; "Bush Administration Proposes Draft Regulation Gutting Protections for Nation's Endangered Species," news release, Center for Biological Diversity, Web site, Aug. 11, 2008; "An Endangered Act," editorial, *New York Times*, Aug. 13, 2008, 22. For the proposed regulation, see "Sensitive Species and Endangered Species Act Section 7 Conservation Policy for National Forest System Land Management Planning under the 2008 Planning Rule," *Federal Register*, Aug. 8, 2008.

13. Some implications of this "tangle of authorities" were reflected in a decision of the 11th U.S. Circuit Court of Appeals in which the panel rejected the argument of the Forest Service that the issue before the court was moot on the ground that the court could not determine whether the regulations in question "have been abandoned or still govern a number of logging projects." See *Ouachita Watch League v. Jacobs* (Sept. 6, 2006) and related news item by the AP, "Circuit Court Finds Environmental Groups Had Standing to Sue Forest Service" in AccessNorthGa.com, Sept. 22, 2006.

14. Wolf, "National Forest Timber Sales and the Legacy of Gifford Pinchot."

SOURCES

The works listed below represent the extensive scholarship devoted to Forest Service history since 1970. As the most readily available research materials, they provided the groundwork for this revisionist history of the agency and would be profitable resources for readers to consult.

Alverson, William S., Walter Kuhlmann, and Donald M. Waller. *Wild Forests: Conservation Biology and Public Policy.* Washington, DC: Island Press, 1994.
Baldwin, Donald N. *The Quiet Revolution: The Grass Roots of Today's Wilderness Preservation Movement.* Boulder, CO: Pruett, 1972.
Barker, Rocky. *Scorched Earth: How the Fires of Yellowstone Changed America.* Washington, DC: Island Press, 2005.
Barron, Jill S., ed. *Rocky Mountain Futures: An Ecological Perspective.* Washington, DC: Island Press, 2002.
Bolgiano, Chris. *The Appalachian Forest.* Mechanicsburg, PA: Stackpole, 1998.
Botti, William B., and Michael D. Moore. *Michigan's State Forests: A Century of Stewardship.* East Lansing: Michigan State University Press, 2006.
Boyd, William, and Scott Prudham. "Manufacturing Green Gold: Industrial Tree Improvement and the Power of Heredity in the Postwar United States." In *Industrial Organisms: Introducing Evolutionary History,* ed. Susan R. Schrepfer and Philip Scranton, 107–39. New York: Routledge, 2004.
Burgess, Bonnie B. *Fate of the Wild: The Endangered Species Act and the Future of Biodiversity.* Athens: University of Georgia Press, 2001.
Cassidy, Victor M. *Henry Cowles, Pioneer Ecologist.* Chicago: Sigel Press, 2007.
Catton, Theodore, and Lisa Mighetto. *The Fish and Wildlife Job on the National Forests: A Century of Game and Fish Conservation, Habitat Protection, and Ecosystem Management.* Washington, DC: Historical Research Associates, prepared for the U.S. Forest Service, October 1998.
Clary, David A. *Timber and the Forest Service.* Lawrence: University Press of Kansas, 1986.
Clepper, Henry. *Professional Forestry in the United States.* Washington, DC: Resources for the Future, 1971.
DeCoster, Lester A. *The Legacy of Penn's Woods, 1895–1995: A History of the Pennsylvania Bureau of Forestry.* Harrisburg: Pennsylvania Historical and Museum

Commission and Pennsylvania Department of Conservation and Natural Resources, 1995.
Devall, Bill, ed. *Clearcut: The Tragedy of Industrial Forestry*. San Francisco: Sierra Club Books and Foundation for Deep Ecology, 1994.
Dombeck, Michael P., Christopher A. Wood, and Jack E. Williams. *From Conquest to Conservation: Our Public Lands Legacy*. Washington, DC: Island Press, 2003.
Drummond, Alexander. *Enos Mills: Citizen of Nature*. Boulder: University Press of Colorado, 1995.
Duane, Timothy P. *Shaping the Sierra: Nature, Culture, and Conflict in the Changing West*. Berkeley: University of California Press, 1999.
Dunlap, Thomas R. *Saving America's Wildlife: Ecology and the American Mind, 1850–1990*. Princeton: Princeton University Press, 1988.
Durbin, Kathie. *Tongass: Pulp Politics and the Fight for the Alaska Rain Forest*. Corvallis: Oregon State University Press, 1999.
Forrest, Suzanne S. "The Vallecitos Federal Sustained-Yield Unit: The (All Too) Human Dimension of Forest Management in Northern New Mexico, 1945–1998." In *Forests under Fire: A Century of Ecosystem Mismanagement in the Southwest*, ed. Christopher J. Huggard and Arthur R. Gómez, 67–92. Tucson: University of Arizona Press, 2001.
Frome, Michael. *The Forest Service*. 2nd ed. Boulder, CO: Westview Press, 1984.
Frye, Bob. *Deer Wars: Science, Tradition, and the Battle over Managing Whitetails in Pennsylvania*. University Park: Pennsylvania State University Press, 2006.
Gilligan, James P. "The Development of Policy and Administration of Forest Service Primitive and Wilderness Areas in the Western United States." 2 vols. PhD dissertation, University of Michigan, 1953.
Glover, James M. *A Wilderness Original: The Life of Bob Marshall*. Seattle: Mountaineers, 1986.
Graham, Frank, Jr. *The Adirondack Park: A Political History*. Syracuse, NY: Syracuse University Press, 1978.
Hadley, Diana. "Grazing the Southwest Borderlands." In *Forests under Fire: A Century of Ecosystem Management in the Southwest*, ed. Christopher J. Huggard and Arthur R. Gómez, 93–131. Tucson: University of Arizona Press, 2001.
Harvey, Mark. *Wilderness Forever: Howard Zahniser and the Path to the Wilderness Act*. Seattle: University of Washington Press, 2005.
Hays, Samuel P. "A Challenge to the Profession of Forestry." In *Practicing Stewardship and Living a Land Ethic: Proceedings of the 1991 Penn State Forest Resources Issues Conference*, ed. James C. Finley and Stephen B. Jones, Harrisburg, Pa., March 26–27, 1991, Pennsylvania State University, 1992.
———. *Conservation and the Gospel of Efficiency*. Cambridge, MA: Harvard University Press, 1959; reprint, Pittsburgh: University of Pittsburgh Press, 1999.

———. *Forest Papers*. 3 vols. Boulder, CO: privately printed, 2003.
———. "Human Choice in the Great Lakes Wildlands." In *Environmental Change in the Great Lakes Forest*, ed. Susan Flader, 295–318. Minneapolis: University of Minnesota Press, 1982.
———. "The New Environmental Forest." *University of Colorado Law Review* 59 (1988): 517–50.
———. *Wars in the Woods: The Rise of Ecological Forestry in the United States*. Pittsburgh: University of Pittsburgh Press. 2006.
Hirt, Paul. *A Conspiracy of Optimism: Management of the National Forests since World War II*. Lincoln: University of Nebraska Press, 1994.
Houck, Oliver A. *The Clean Water Act TMDL Program: Law, Policy, and Implementation*. Washington, DC: Environmental Law Institute, 1999.
Huggard, Christopher J. "America's First Wilderness Area: Aldo Leopold, the Forest Service, and the Gila of New Mexico, 1924–1980." In *Forests under Fire: A Century of Ecosystem Management in the Southwest*, ed. Christopher J. Huggard and Arthur R. Gómez, 133–80. Tucson: University of Arizona Press, 2001.
Huggard, Christopher J., and Arthur R. Gómez, eds. *Forests under Fire: A Century of Ecosystem Mismanagement in the Southwest*. Tucson: University of Arizona Press, 2001.
Jenkins, Jerry, et al. *Acid Rain in the Adirondacks: An Environmental History*. Ithaca, NY: Cornell University Press, 2007.
Josephson, Paul R. *Motorized Obsessions: Life, Liberty, and the Small-Bore Engine*. Baltimore: Johns Hopkins University Press, 2007.
Joslin, Les. *The Wilderness Concept and the Three Sisters Wilderness*. Bend, OR: Wilderness Associates, 2000.
Kaufman, Herbert. *The Forest Ranger: A Study in Administrative Behavior*. 1960. Reprint, Washington, DC: Resources for the Future, 2006.
Kaufmann, Harold F., and Lois C. Kaufmann. "Toward the Stabilization and Enrichment of a Forest Community." In *Communities and Forests: Where People Meet the Land*, ed. Robert G. Lee and Donald R. Field, 96–111. Corvallis: Oregon State University Press, 2005.
Kempton, Willett, James S. Boster, and Jennifer A. Hartley. *Environmental Values in American Culture*. Cambridge, MA: MIT Press, 1995.
Klyza, Christopher McGrory, and Stephen C. Trombulak, eds. *The Future of the Northern Forest*. Hanover, NH: University Press of New England, 1994.
Knight, Richard L., and Sarah F. Bates, eds. *A New Century for Natural Resources Management*. Washington, DC: Island Press, 1995.
Koontz, Tomas M. *Federalism in the Forest: National versus State Natural Resource Policy*. Washington, DC: Georgetown University Press, 2002.

Langston, Nancy. *Forest Dreams, Forest Nightmares: The Paradox of Old Growth in the Inland West*. Seattle: University of Washington Press, 1995.

LeMaster, Dennis C. *Decade of Change: The Remaking of Forest Service Statutory Authority during the 1970s*. Westport, CT: Greenwood Press, 1984.

Lindstrom, Matthew J., and Zachary A. Smith. *The National Environmental Policy Act: Judicial Misconstruction, Legislative Indifference, and Executive Neglect*. College Station: Texas A&M University Press, 2001.

Lockmann, Ronald F. *Guarding the Forests of Southern California: Evolving Attitudes toward Conservation of Watershed, Woodlands, and Wilderness*. Glendale, CA: A. H. Clark, 1981.

Luoma, Jon R. *The Hidden Forest: The Biography of an Ecosystem*. Corvallis: Oregon State University Press, 2006.

Maher, Neil M. *Nature's New Deal: The Civilian Conservation Corps and the Roots of the American Environmental Movement*. New York: Oxford University Press, 2008.

Marsh, Kevin R. *Drawing Lines in the Forest: Creating Wilderness Areas in the Pacific Northwest*. Seattle: University of Washington Press, 2007.

Marshall, Suzanne. *"Lord, We're Just Trying to Save Your Water": Environmental Activism and Dissent in the Appalachian South*. Gainesville: University Press of Florida, 2002.

Matthews, Mark. *Smoke Jumping on the Western Fire Line: Conscientious Objectors during World War II*. Norman: University of Oklahoma Press, 2006.

McCloskey, Michael. *In the Thick of It: My Life in the Sierra Club*. Washington, DC: Island Press, 2005.

McMartin, Barbara. *The Great Forest of the Adirondacks*. Utica, NY: North Country Books, 1994.

Meine, Curt. *Aldo Leopold: His Life and Work*. Madison: University of Wisconsin Press, 1988.

Miller, Char, ed. *American Forests: Nature, Culture, and Politics*. Lawrence: University Press of Kansas, 1997.

———. *Gifford Pinchot and the Making of Modern Environmentalism*. Washington, DC: Island Press, 2001.

Morrison, Ernest, J. *Horace McFarland: A Thorn for Beauty*. Harrisburg: Pennsylvania Historical and Museum Commission, 1995.

Moul, Francis. *The National Grasslands: A Guide to America's Undiscovered Treasures*. Lincoln: University of Nebraska Press, 2006.

Nash, Roderick Frazier. *Wilderness and the American Mind*. 4th ed. New Haven: Yale University Press, 2001.

Nie, Martin. *The Governance of Western Public Lands: Mapping Its Present and Future*. Lawrence: University Press of Kansas, 2008.

O'Toole, Randal. *Reforming the Forest Service*. Washington, DC: Island Press, 1988.

Peterson, Shannon. *Acting for Endangered Species: The Statutory Ark.* Lawrence: University Press of Kansas, 2002.
Pinchot, Gifford. *Breaking New Ground.* 1947. Commemorative ed., Washington, DC: Island Press, 1998.
Pinkett, Harold T. *Gifford Pinchot: Private and Public Forester.* Urbana: University of Illinois Press, 1970.
Proescholdt, Kevin, Rip Rapson, and Miron L. Heinselman. *Troubled Waters: The Fight for the Boundary Waters Canoe Area Wilderness.* St. Cloud, MN: North Star Press of St. Cloud, 1995.
Pyne, Stephen J. *Fire in America: A Cultural History of Wildland and Rural Fire.* Princeton: Princeton University Press, 1982.
Reiger, John F. *American Sportsmen and the Origins of Conservation.* 3rd ed. Corvallis: Oregon State University Press, 2001.
Righter, Robert W. *The Battle over Hetch Hetchy: America's Most Controversial Dam and the Birth of Modern Environmentalism.* New York: Oxford University Press, 2006.
Robbins, William G. *American Forestry: A History of State and Private Cooperation.* Lincoln: University of Nebraska Press, 1985.
Robinson, Gordon. *The Forest and the Trees: A Guide to Excellent Forestry.* Washington, DC: Island Press, 1988.
Rodgers, Andrew Denny, III. *Bernhard Eduard Fernow: A Story of North American Forestry.* Durham, NC: Forest History Society, 1991.
Ross, Ken. *Environmental Conflict in Alaska.* Boulder: University Press of Colorado, 2000.
Roth, Dennis M. *The Wilderness Movement and the National Forests.* College Station, TX: Intaglio Press, 1988.
Rothman, Hal. *Preserving Different Pasts: The American National Monuments.* Urbana: University of Illinois Press, 1989.
———. "'A Regular Ding-Dong Fight': The Dynamics of Park Service–Forest Service Controversy during the 1920s and 1930s." In *American Forests: Nature, Culture, and Politics,* ed. Char Miller, 109–24. Lawrence: University Press of Kansas, 1997.
Rowley, William D. *U.S. Forest Service Grazing and Rangelands: A History.* College Station: Texas A&M University Press, 1985.
Schneider, Paul. *The Adirondacks: A History of America's First Wilderness.* New York: Henry Holt, 1997.
Schrepfer, Susan R. "Establishing Administrative 'Standing': The Sierra Club and the Forest Service, 1897–1956." In *American Forests: Nature, Culture and Politics,* ed. Char Miller, 125–42. Lawrence: University Press of Kansas, 1997.
———. *The Fight to Save the Redwoods: A History of Environmental Reform, 1917–1978.* Madison: University of Wisconsin Press, 1983.

———. *Nature's Altars: Mountains, Gender, and American Environmentalism.* Lawrence: University Press of Kansas, 2005.
Scott, Doug. *The Enduring Wilderness: Protecting Our Natural Heritage through the Wilderness Act.* Golden, CO: Fulcrum, 2005.
Sedjo, Roger A., ed. *A Vision for the Forest Service: Goals for Its Next Century.* Washington, DC: Resources for the Future, 2000.
Sellers, Richard West. *Preserving Nature in the National Parks: A History.* New Haven: Yale University Press, 1997.
Servid, Carolyn, and Donald Snow, eds. *The Book of the Tongass.* Minneapolis: Milkweed Editions, 1999.
Smith, Thomas G. *Green Republican: John Saylor and the Preservation of America's Wilderness.* Pittsburgh: University of Pittsburgh Press, 2006.
Smyth, Arthur V. *Millicoma: Biography of a Pacific Northwestern Forest.* Durham, NC: Forest History Society, 2000.
Steen, Harold K, ed. *Forest and Wildlife Science in America: A History.* Durham, NC: Forest History Society, 1999.
———. *Forest Service Research: Finding Answers to Conservation's Questions.* Durham, NC: Forest History Society, 1998.
———. *Jack Ward Thomas: The Journals of a Forest Service Chief.* Durham, NC: Forest History Society, 2004.
———. *The U.S. Forest Service: A History.* Seattle: University of Washington Press, 1976.
Sutter, Paul S. *Driven Wild: How the Fight against Automobiles Launched the Modern Wilderness Movement.* Seattle: University of Washington Press, 2002.
Tober, James. *Who Owns the Wildlife? The Political Economy of Conservation in Nineteenth-Century America.* Westport, CT: Greenwood Press, 1981.
———. *Wildlife and the Public Interest: Nonprofit Organizations and Federal Wildlife Policy.* New York: Praeger, 1989.
Travis, William R. *New Geographies of the American West.* Washington, DC: Island Press, 2007.
Trefethen, James B. *An American Crusade for Wildlife.* New York: Boone and Crockett Club, 1975.
Twight, Ben W. *Organizational Values and Political Power: The Forest Service versus the Olympic National Park.* University Park: Pennsylvania State University Press, 1983.
Vaughn, Jacqueline, and Hanna J. Cortner. *George W. Bush's Healthy Forests: Reframing the Environmental Debate.* Boulder: University Press of Colorado, 2005.
Warren, Louis S. *The Hunter's Game: Poachers and Conservationists in Twentieth-Century America.* New Haven: Yale University Press, 1997.

Way, Albert G. "Burned to Be Wild: Herbert Stoddard and the Roots of Biological Conservation in the Southern Longleaf Pine Forest." *Environmental History* 11, no. 3 (July 2006): 500–526.

The Wilderness Society et al. *National Forest Planning: A Conservationist's Guide.* 2nd ed. Washington, DC: Wilderness Society, 1983.

Wilkinson, Charles F., and H. Michael Anderson. *Land and Resource Planning in the National Forests.* Washington, DC: Island Press, 1987.

Williams, Michael. *Americans and Their Forests: A Historical Geography.* Cambridge: Cambridge University Press, 1989.

Wolf, Robert E. "National Forest Timber Sales and the Legacy of Gifford Pinchot: Managing a Forest and Marking It Pay." In *American Forests: Nature, Culture, and Politics,* ed. Char Miller, 87–105. Lawrence: University Press of Kansas, 1997.

Wolf, Tom. *Arthur Carhart: Wilderness Prophet.* Boulder: University Press of Colorado, 2008.

Yaffee, Steven Lewis. *The Wisdom of the Spotted Owl: Policy Lessons for a New Century.* Washington, DC: Island Press, 1994.

Yahner, Richard H. *Eastern Deciduous Forest Ecology and Wildlife Conservation.* Minneapolis: University of Minnesota Press, 1995.

INDEX

abandoned lands, 3
accounting, 97–100, 129–30, 178n50
acid rain, 115
Adams, Charles Kendall, 154n24, 165n50
adaptive management, 23, 152n59
Adirondack Park, New York, 26–27, 31, 42, 53, 153n6
administrative choices. *See* decision making
Advertising Council, 102
aesthetic corridors, 42, 84, 85, 168n84
aesthetics: Adirondack Park and, 26–27; agency approach to, 41–45, 83–85; clear-cutting and, 83–85; conflicting opinions on, 13–14, 42, 83–84; and ecology, 107, 110; non-agency venues for, 50–51; parks and, 44–45, 83–84, 157n47; public appreciation of, 41–42, 50, 92; recreation and, 72–73; wilderness protection and, 85–86, 91
Alaska, 61–62
Alaska Native Claims Settlement Act (1971), 62
Alaska Pulp Corporation, 61
Aldo Leopold Wilderness Center, 91
Aldo Leopold Wilderness Research Institute, 151n43
allowable cut, 10, 63, 96, 162n17
American Civic Association, 44
American Conservation (magazine), 51
American Forest and Paper Association, 58
American Forest Congress, 160n78
American Forest Institute, 108
American Forestry Association, 49–51, 147n15, 160n75, 166n53; *Forestry and Irrigation*, 149n29
American Institute of Biological Sciences, 109
American Society of Environmental History, ix
American Society of Landscape Architects, 88
antipredator campaigns, 37, 76
Antiquities Act (1906), 49
Apache National Forest, 38
Appalachian Trail, 165n46
appeal system, 127–28

Aquatic Conservation Strategy, 179n59
area management, 65
Arthur Carhart Wilderness Training Center, 91, 151n43
Aspinall, Wayne, 94, 166n53
automobiles: and access, 13, 71, 74, 103; and aesthetics, 83–84; and camping, 13, 71–72, 165n41; increase in numbers of, 165n42; and wildlife, 13, 74

beauty strips, 65, 84, 85
best management practices (BMPs), 117, 124, 132
Biltmore estate, North Carolina, 46, 53, 151n52
biodiversity, 110–13, 124
bison, 34, 35, 49
Bitterroot National Forest, 121
Black, Frank, 53
BMPs. *See* best management practices
Boone and Crockett Club, 13, 34, 36, 74
Bosworth, Dale, 11, 22, 133–34, 138
botany, 16
Boundary Waters Canoe Area, Minnesota, 70, 75, 88
Brady, John G., 52
Brannan, Charles F., 163n30
Bridger-Teton National Forest, 141
Brower, David, 101
Brush Disposal Fund, 98
budget. *See* finances, of forest management
Bureau of Forestry, 27, 29, 44, 46, 53
Bureau of Land Management, 5, 91
Bush, George W., 5–6, 19, 117–18, 128, 133–36, 138–39, 142–43, 176n27, 177n36, 177n40

California, 117
Cannon, Joseph, 158n54
car camping, 13, 71–72, 165n41. *See also* automobiles
Carhart, Arthur, 69–70, 85–88
Carson National Forest, 104

191

Catton, Theodore, 155n32
Chamberlain-Hayden bill, 155n35
Cherrington, Charles, 167n78
Chesapeake Bay, 141
Chicago Wilderness, 173n4
Chippewa Forest Reserve, 53–54
citizen groups. *See* public
Civilian Conservation Corps (CCC), 20, 60, 72, 89
Clarke, John D., 148n18
Clarke-McNary Act (1924), 9, 60, 148n18
Clary, David, 29, 82, 162n19
Clawson, Marion, 157n46
Clean Air Act (1970), 123–24, 127
Clean Water Act (1972), 123–24, 127, 179n57
clear-cutting, xii, 65–66, 85
Clepper, Henry, 35, 39, 46, 51, 64, 148n18, 155n32, 156n39, 161n5, 161n10
Cliff, Edward, 95, 150n39, 163n20
Clinton, Bill, 14, 117, 133–34, 138–39, 175n26
Congressional Budget Office (CBO), 128
Congressional Research Service (CRS), 129
conservation. *See* forest management; wildlife management
conservation easements, 11
controlled burns, 122
Cooperative Sustained-Yield Program, 61
Cornell University, 84
corridors. *See* aesthetic corridors
Crowell, John, 96
cut-over land, 3, 8, 57, 60

Darling, Jay "Ding," 77
de facto wilderness areas, 89–90
DeBonis, Jeff, 151n46
decision making, 48–52; by executive branch, 48–49; experts and, 50; in internal affairs, xi; patterns in, xii; society-agency relationships and, xi–xii, 145n5
deer, 35–36, 76, 87
dendrology, 16
Denver Public Lands Convention (1907), xii
disturbance, 118–22
diversity of species, 124
Dombeck, Michael, 133–34, 139, 172n124
Drummond, Alexander, 45
drylands, 146n2

earned harvest effect, 10, 162n17
easements. *See* conservation easements

ecological forestry: challenges presented by, 141; and disturbance issues, 118–22; implementation of, 122–26; legacy of, 135–36; opposition to, 133–35, 142–43; overview of, 111–12; and procedural issues, 132–33; and restoration issues, 118–22; and species, 112–14, 120, 131–32; and watershed management, 114–18; wood production vs., 121–22, 131, 141
Ecological Society of America, 109
ecology: aesthetics and, 107, 110; and biodiversity, 110–13, 124; emergence of, 106; and fire, 121; forest management and, 23, 108–11, 141; forest objectives based on, 122–26; and habitats, 113–14, 118–22, 142; problems highlighted by, 120–21; as public concern, 14, 107–10, 119–20, 125–28; as scientific concern, 16–19, 107, 109–11, 119–20, 125. *See also* ecological forestry
education. *See* forestry schools
endangered species, 152n61
Endangered Species Act (ESA, 1973), 106, 108, 112, 115, 123, 127, 133, 134, 142–43
enforcement, 21–22
environmental community, 126, 163n24
environmental issues. *See* ecological forestry; ecology
Environmental Protection Agency (EPA), 83, 95, 115, 117, 124, 132, 136, 175n21, 179n57
EPA. *See* Environmental Protection Agency
ESA. *See* Endangered Species Act (ESA, 1973)
European forest management, 7, 39
even-age management, 65
Everett, D. C., 64
experts: agency decision making and, 50; ecological, 110, 125; nontraditional, 18–19, 125–26, 151n46; in public realm, 15

Federal Aid Road Act (1916), 88, 165n42
Federation of Women's Clubs, Minnesota, 53
Fernow, Bernhard, 27, 32, 42, 53, 84, 154n24, 157n47, 168n84
finances, of forest management, 41, 46–48, 69, 97–100, 128–30, 178n50
Finley, Jim, 165n51
fire: cut-over land and, 3; ecological forestry and, 121; importance of, 20; labor for fighting, 20; media attention to, 121–22; policies on, 9, 20, 60; and public support, 101–2; states and, 60, 148n18

fire ecology, 78
fire prevention, 20–21
fire suppression, 20–21
fish and wildlife programs, 13
Flower, Roswell, 53
Forest and Stream (magazine), 34
forest cover, and streamflow, 33
forest denudation, 31
forest experiment stations, 17
forest fragmentation, 120
forest health, 97, 121–22, 176n31
Forest Homestead Act (1906), 4
forest industry: Forest Service relationship with, 8–11, 39–40, 56–63, 131, 162n19; long-term interests of, 58; regulation of, 8–9, 57–61, 147nn13–15; state relationships with, 9
Forest Legacy Program, 11
forest management: adaptive, 23, 152n59; of Adirondack Park, 26–27; and aesthetics, 42–45; area vs. single-tree, 65; capacity for, 19–20; conflicting opinions on, 16–17, 26, 104; developments in, 19–23; ecological forestry, 111–26, 131–36, 141–43; and ecology, 23, 108–11, 141; and enforcement, 21–22; finances and, 41, 46–48, 69, 97–100; fire and, 19–21; industrial forestry, 63–66; market influences on, 10–11, 39–40, 58–59, 62, 104; Pinchot and, 26–27, 35–36, 42–48, 52–54, 97, 155n32, 158n54; recreational uses and, 22–23; states and, 9; sustained-yield, 8, 10, 38, 59, 61; and wildlife, 94. *See also* wildlife management
Forest Management Act (1897), xi, xii, 25, 27–28, 30–32, 65, 91
forest professionals, development of, 15–19
forest reserves: conflicting opinions on, 3–4, 26; establishment of, 1–3, 25, 34, 149n28; legislation on, 159n71; policies on, 2–3; public attitudes toward, 12; sportsmen and, 34, 50, 74; supporters of, 43–44; as wildlife refuges, 35
forest science: conflicting opinions on, 19; developments in, 15–19, 109, 150n37; and industrial forestry, 63–66; specialists and, 18–19, 151n46
Forest Service Employees for Environmental Ethics (FSEEE), 140
forestry schools, 17, 150n41
FORPLAN computer program, 96

Franklin, Jerry, 174n12
Fremont Experiment Station, 33
"further research" agendas, 33
future generations, 109

game preservation. *See* wildlife management
gap analysis, 171n114
Geer v. Connecticut, 77
General Accounting Office, 99, 129
General Federation of Women's Clubs, 44–45
Giant Sequoia National Monument, 138
Gifford Pinchot National Forest, 82
Gila National Forest, 86, 87
Gilligan, James, 89
Governors' Conservation Conference (1908), 13, 44–45, 50–51
Grand Canyon Forest Reserve, 44, 149n27
Grand Canyon National Game Preserve, 35
Granger, Christopher M., 162n20
Graves, Henry, 17, 48, 58, 157n53, 158n55, 158n58
grazing: as controversial issue, 5, 28–30, 146n5; importance of, xi, 12–13, 28–30; management of, 27–30, 66–68, 153n15; market for, 66–68; in national forests, 68, 164n32; permits for, 29, 67–68, 163n28; rights to, 68, 164n31; watershed management and, 80–81; wildlife management and, 36–37, 76
Great Smoky Mountains National Park, 84, 113
Greater Yellowstone Area Clean Air Partnership, 141
The Greatest Good (DVD), 102, 145n6, 150n35, 151n46, 174n15
Greeley, William, 8, 58, 86, 98, 146n5
Grinnell, George Bird, 34

H. J. Andrews Experimental Forest, 174n12
habitats: classification of, 114; disturbance and restoration of, 118–22; ecological forestry and, 114, 142; forest management and, 77, 94, 113–14; geographic extent of, 114, 170n114
Hall, Ansel, 165n50
Harlow, William, 16
Harrar, Ellwood, 16
harvests, watershed effects of, 117
Herbert, Paul, 150n41
Hetch Hetchy dam, xii, 43

Hill, David, 53
Hirt, Paul, 63, 71, 82, 165n41
homesites, 69–70
homestead settlement, 4–5
Hoover, Herbert, 101
Hunt v. United States, 77
hunting. *See* sportsmen
Hutchinson, Wallace, 69
hydrology, 23, 116

industrial forestry, 63–66
irrigation, 32, 149n29
Izaak Walton League, 75, 88

Jones, Steve, 165n51
Journal of Forestry, 78, 126

Kaibab Forest Game Reserve, 76
Keep Commission, 157n53
Ketchikan Pulp Company, 61
Kirtland's warbler, 142
Kneipp, Leon, 36, 75
Knutson-Vandenburg Act (1930), 98

labor: and cost reduction, 64–65; mill jobs, 40, 59, 64, 156n44
Lacey Act (1900), 49
Lacey, John, 44, 49, 157n53, 159n72
landscape architects, 84, 85
landscape management areas, 85
Langston, Nancy, 38, 150n37, 155n28, 156n41, 156n42, 156n45, 161n6, 161n7, 163n25
Leopold, Aldo, 35–38, 69, 75, 79–81, 86, 87, 104–5, 168n92, 169n94, 169n95
Lewis, James G., *The Forest Service and the Greatest Good*, ix–x
Light v. United States, 25
"limits of acceptable change," 91
Lincoln-Scapegoat area, Montana, 94, 166n53
Lingua Botanica (newsletter), 172n2
livestock. *See* grazing
L-20 regulation, 87, 88, 169n93
lumber mills, 40, 59, 156n43

"making the forests pay," 41, 46–48, 69, 97, 130
management. *See* forest management; planning; public land management; watershed management; wildlife management
management indicator species (MIS), 132, 141, 179n58

Manring, Nancy, 126
marginal lands provision, 157n46
market: forest management influenced by, 10–11, 39–40, 58–59, 62, 104; grazing and, 66–68; wood production and, 57, 66–67
Marsh, George Perkins, *Man and Nature*, 31, 154n21
Marshall, Bob, 86, 168n92
Mason, Ira J., 29, 148n20, 156n44
Massabesic Experimental Forest, 113
maximum production, 10, 62–63, 97, 148n20, 148n21
McArdle, Richard, 8, 58, 89
McCloskey, Mike, 166n53
McNary-McSweeney Act (1929), 17
media: fire safety and, 101–2; and forest health, 121–22
Meine, Curt, 37, 80
Merton, Robert, 143
Mighetto, Lisa, 155n32
mills. *See* lumber mills
Mills, Enos, 45
Minckler, Leon, 16
Mission 66, 73
Monongahela clear-cutting case, xii, 66, 163n24
Morris, Robert P. W., 53
motorized recreation vehicles, 93
Mountain States Legal Foundation, 5
Muir, John, 45
Multiple-Use and Sustained-Yield Act (1960), 14, 30, 43, 54, 56, 90–92, 94, 99, 103, 105, 108, 111, 123, 139–40
Municipal Watershed Act (1940), xi, 82

National Academy of Sciences, 32, 44, 149n27
National Association of Forestry Schools and Colleges, 19
National Association of State Foresters, 19
National Conservation Association, 51
National Environmental Policy Act (NEPA, 1969), 18, 96, 106, 108, 123, 134, 135
National Forest Management Act (NFMA, 1976), 14, 90, 92, 100, 106, 108, 113, 124, 127, 134, 135
national forests: establishment of, 7, 27; grazing in, 68; homesteading in, 4–5; national role of, 11; parks and, 44; Pinchot and, 13; public and, 49–51, 107–9; role of, 106–9; uses of, 12–15, 21–22, 52–53, 92–97, 99–100, 103–5, 139–40

national grasslands, 146n2
National Irrigation Congress, 160n75
National Lumber Manufacturers Association, 57, 58, 99
national parks: and aesthetics, 44–45, 83–84, 157n47; timber harvest in, 44, 158n58
National Parks and Conservation Association, 16
National Park Service: establishment of, 13, 45, 158n55; Forest Service vs., 73–74, 84, 123, 138, 158n55; interpretation activities of, 73, 102, 165n50; popular support for, 102; and recreation, 73–74; and wilderness, 91; and wildlife management, 113
National Trails Act, 72
National Wild and Scenic Rivers Act, 72
National Wild and Scenic Rivers System, 85
National Wilderness Preservation Act (1964), 89
National Wildlife Federation, 77
National Woolgrowers Association, 154n15
Natural Areas Association, 109
Natural Resources Defense Council, 129
natural world. *See* ecology
NEPA. *See* National Environmental Policy Act (NEPA, 1969)
New Mexico Game Protective Association, 75
New York. *See* Adirondack Park, New York
Newell, Frederick, 160n78
NFMA. *See* National Forest Management Act (NFMA, 1976)
Non-Industrial Private Forests (NIPF), 91
North American Wildlife Policy, 167n69
North Cascades National Park, 85
North Central Experiment Station, 91
Northwest forest plan, 117–18, 175n26, 176n27, 179n56

off-road vehicles, 22
old-growth forests, 7, 38–39, 110
Olympic National Park, 82, 85
Olympic Peninsula, 82
Operation Outdoors, 73
O'Toole, Randy, 96, 129
Outdoor Art League of California, 158n54
outdoor recreation. *See* recreation
outdoor recreation industry, 72–73, 165n47

Pacific Northwest Aquatic Conservation Strategy, 179n59

Pacific Northwest forest plan, 117–18, 175n26, 176n27, 179n56
pack trips, 166n53
parks. *See* national parks
peer review, 134, 136
Pennsylvania, 31
permit-and-fee system, 29, 67–68, 163n28
Pinchot, Gifford: as administrator, 49–51, 100; and aesthetic issues, 42–45; and associations, 147n15, 160n78; and Biltmore estate, 46, 151n52; *Breaking New Ground*, ix–x, 53; controversies involving, xii, 5; and finances, 46–47, 98, 130; and fire, 21; and forest management, 26–27, 35–36, 42–48, 52–54, 97, 155n32, 158n54; on forest science, 150n37; as governor, 147n14; and grazing, 29–30, 32; legacy of, 54; and national parks, 44, 158n55; objectives emphasized by, 13, 25–26, 137–38; and private industry, 8–9, 57, 147nn13–15; and wildlife, 35, 74
Pinchot-Ballinger controversy, xii
Pine Cone (newsletter), 104
plains states, 146n2
planning: Bush administration policy revision on, 134, 136, 177n36, 177n40; and ecology, 96–97, 110, 111, 124–25, 134–36; for forest uses, 92–93, 95–97, 100; public commentary on, 124–25
Plum Creek Timber Company, 11
pollution, 115
Port Angeles, Washington, 82
Potter, Albert, 29–30, 153n15, 154n20
predators, campaigns against, 37, 76
Prescott National Forest, 80
primitive areas, 88
private forest industry. *See* forest industry
private land ownership, policy shift away from, 2
privatization theory, 5–6
production forestry, 17
Property and Environment Research Center (PERC), 6
public: and budget, 128–30; and ecology, 14, 107–10, 119–20, 125–28; fire safety and, 101–2; and Forest Service, 49–50, 100–103, 126–30; forest use by, 92; knowledgeability of, 126; and national forests, 49–51, 107–9; and National Park Service, 102; and recreation, 71–74; and wilderness, 89–90, 94; and wildlife, 77

INDEX   195

public benefits, 98
public land management, 4–7, 146n1, 146n2
public land policy: preliminary to U.S. Forest Service, 1–3; recent public opinion on, 6–7; Republican Party and, 5–6
Public Lands Committee, 49
public lands, rights to, 3–4, 28, 146n5
pulp and paper industry, 11, 58, 64

radio telemetry, 94, 112, 114, 174n20
Rainier Forest Reserve, 44, 149n27
range lands. *See* grazing
Reagan, Ronald, 5–6, 96, 133, 138
real estate development, 11, 21
Reclamation Service, 50
recreation: and aesthetics, 72–73; agency approach to, 69–74; conflicting opinions on, 70; diversity of, 22; emergence of, 17; income from, 93; management of, 22–23; motorized vehicles for, 93; National Park Service and, 73–74; and wilderness, 93–94; World War II's effect on, 56
redwoods, 138
regulated forests, 7–8, 38–39, 156n39
regulation of forest industry, 8–9, 57–61, 147nn13–15
Reid, T. R., 147n11
Reiger, John, 149n28, 159n72
replanting, 60
Replanting Fund, 98
Republican Party, 5, 133–34
research: agency and, 17, 33; forest industry and, 64
Resources for the Future, 157n46
restoration, 118–22
rights: grazing, 68, 164n31; to public lands, 3–4, 28, 146n5
Rio Grande National Forest, 33
road construction, 59, 61, 86–89, 119, 121, 128, 130, 138–39. *See also* aesthetic corridors
Roads and Trails Fund, 98
Rocky Mountain National Park, 45
Rodgers, Andrew, 53
Roosevelt, Franklin D., 60, 77, 82
Roosevelt, Theodore, 4, 13, 34, 35, 49, 50, 53, 54, 61, 100, 155n32, 157n53, 158n54, 160n78
Roth, Filibert, 29–30, 154n18
Rowley, William D., *U.S. Forest Service Grazing and Rangelands*, 146n5, 163n28, 164n32

Sagebrush Rebellion, 6
salmon, 95, 117, 131–32, 175n26
Salvage Sale Fund, 98
San Isabel National Forest, 70
Sargent, Charles, 26, 153n6
science. *See* ecology; forest science; silviculture
Science Advisory Committee, 14
Scott, Michael, 171n114
secret forests, 107, 110
Sedjo, Roger, 147n7
seeds, 64
sensitive species, 124
Sequoia Forest Reserve, 44
Sequoia National Forest, 138
Sequoia National Park, 44
sheep grazing, and watershed protection, 32
Shenandoah National Park, 84
Shipstead-Nolan Act (1930), 88
Shoshone National Forest, 141
Sierra Club, 84, 100–101
Sierra Nevada region, California, 176n29
Silcox, Ferdinand, 58
silviculture, 15–16, 38–41
Simpson Lumber Company, 61
Siskiyou National Forest, 121
slash, 21
Smith, Herbert, 168n84
Smokey Bear, 21, 101–2
Society of American Foresters (SAF), 16, 19, 46, 71, 78, 109, 130
society-agency relationships: administrative choices and, xi–xii; changes in, 12–15; significance of, x–xi; trends in, xiii
sociological research, 17
soil erosion, 32, 80–81, 94–95
soil productivity, 41
specialists. *See* experts
species inventories, 113, 131, 141–42, 179n58
species viability, 113, 124, 131–32
sportsmen: and forest reserves, 34, 50, 74; vs. Forest Service, 76; numbers of, 166n54; and wildlife conservation, 33–34, 74–75
spotted owl, 131
sprawl, 11
squatters, 3–4
Stanfield, Robert N., 146n5
State and Private Forestry Division, 60, 91
states: agency-industry relations and, 60–61; and fire control, 60, 148n18; fish and wildlife programs in, 13; forest industry rela-

tionship with, 9; Forest Service relationship with, 77; park and forest administration in, 44, 157n51; regulation by, 9, 61; and wildlife conservation, 75
Steen, Harold, 168n84
stockmen. *See* grazing
stockmen's associations, 67
Stoddard, Herbert, 78
streamflow, 31–33, 154n23, 154n24
Strouder, Deanna J., 177n41
Superior National Forest, 70, 86–88
survey and manage philosophy, 131, 142, 170n114, 174n18, 179n56
sustained-yield management, 8, 10, 38, 59, 61
Swift, Lloyd, 78

technology: wildlife management and, 94, 112, 114, 174n20; wood production and, 64–66
ten o'clock policy, 20
Term Permit Act (1915), 69, 164n34
timber cruising, 65
timber famine, 8, 10, 27, 39, 47, 58
timber mining, 3, 8, 38, 40
timber production. *See* wood production
timber production projections, 8, 10
Timber Sale Program Information Reporting System (TSPIRS), 178n50
Tongass National Forest, 61–62
Tonto National Forest, 81
topography, 40
Transfer Act (1905), 25, 45, 98
transportation, 40–41, 59, 156n45
Trappers Lake, Colorado, 70
travel corridors, 42, 84, 168n84
trust funds, 98–99, 171n124
Turnage, William, 129

U regulations, 87, 88, 168n92
United States v. Grimaud, 25
U.S. Agriculture Department, 25, 28, 34, 48, 54, 157n53, 158n55
U.S. Chamber of Commerce, 64
U.S. Congress: and ecology, 133; and executive branch, 49; and forest reserves, 2, 4; Forest Service and, 5, 45–48, 97–100, 128–30; and national forests, 4–5; and wilderness protection, 89–90, 94, 101, 123; and wildlife management, 78
U.S. Fish and Wildlife Service, 83, 91, 123, 133–34, 136, 142–43, 175n21

U.S. Forest Service: appeals process in, 127–28; autonomy of, 123–24, 177n40; centennial publications of, ix–x, 102, 145n6, 150n35, 151n46; Congress and, 5, 45–48, 97–100, 128–30; constituencies of, 12–15, 50, 103–4; decision making in, xi–xii; Division of Grazing, 37, 76; Division of Range Management, 78; Division of Wildlife, 13, 37, 77, 78; early history of, 25; establishment of, ix, 45; finances of, 46–48, 97–100, 128–30, 178n50; forest industry relationship with, 8–11, 39–40, 56–63, 131, 162n19; future of, 143; legacies of, 52–54, 103–5, 135–43; National Park Service vs., 73–74, 84, 123, 138, 158n55; new vs. old in, 130–35; objectives of, 2, 12–13, 26–27, 52–53, 63, 98, 104, 133–34; opposition to, 146n4; philosophy of, 7; precursors to, 1–4; professionalism of, 52; public's relationship with, 49–50, 100–103, 126–30; scholarship on, ix–x, 25; scientific culture and, 15; state relationships with, 77; timber sales by, 7; Watershed, Fish, and Wildlife Division (WFW), 116, 127, 140, 152n59, 172n2, 177n41
U.S. General Land Office, 29
U.S. Interior Department, 46, 48
U.S. Post Office, 102
U.S. Supreme Court, 25, 29, 77, 179n57
use permits, 21. *See also* permit-and-fee system
uses of national forests: balancing, 92–97; diversity of, 12–15, 21–22, 52–53, 99–100, 103–5, 139–40; problems presented by, 103–4

virgin forests, 7, 10, 90

Wagon Wheel Gap Experiment Station, 33
watershed management: Adirondack Park and, 26–27, 31; agency approach to, xi, 13, 30, 32–33, 50–51, 79–83, 114–18, 132, 155n28; controversies in, 31–33, 175n21; ecological forestry and, 114–18; forest denudation and, 31; grazing and, 80–81; harvests and, 117; interest in, 31–32; Aldo Leopold and, 79–81; problems in, 116–17; soil erosion and, 32, 80–81, 94–95; whole-watershed approach to, 116, 118; wilderness and, 94–95
Watt, James, 5
Watts, Lyle, 58

Waugh, Franklin, *Recreation Use on the National Forests*, 70
Weeks Act (1911), 9, 60, 164n32
West: changing constituencies in, 6; grazing lands in, 28; political involvement of, 4–6; public land use in, 3–4, 6
West Virginia Highlands Conservancy, 176n31
White, Mrs. Lovell, 158n54
Wichita National Forest, 35
Wichita National Wildlife Reserve, 35
wilderness: and aesthetics, 85–86, 91; agency approach to, 85–91, 105, 168n92; citizen groups and, 89–90, 94; Congress and, 89–90, 94, 101, 123; legislation on, 89–90; Aldo Leopold and, 86, 87, 169n94, 169n95; recreation and, 93–94; and watersheds, 94–95; and wildlife, 94; World War II's effect on, 56
Wilderness Act (1964), 87, 93–94, 133
Wilderness Management (manual), 91
Wilderness Society, 129
wildlife: and biodiversity, 110–13; changing perspectives on, 94; non-agency venues for, 50–51, 75; nongame, 170n113; public's interest in, 77; Smokey Bear campaign and, 102. *See also* wildlife management
wildlife management: agency approach to, 74–79, 166n53, 167n69; changing perspectives on, 18; conflicting opinions on, 12, 13, 35–36; and conservation, 33–37; curricula including, 151n41; Division of Wildlife, 13, 37; ecological forestry and, 112–13, 120, 131–32; ecology and, 18; and enforcement, 75; forest science and, 17–18; grazing and, 36–37, 76; Aldo Leopold and, 35–36; states and, 75; World War II's effect on, 56. *See also* habitats; species inventories; species viability; wildlife
Wildlife Management Institute, 77
wildlife reserves, 35
Wildlife Society, 77, 151n44
Willamette National Forest, 171n117, 174n12
Wolf, Robert, 46–48, 129–30, 143
wood production: agency approach to, 12–13, 27, 37–41, 56–63, 95, 137–38, 162n20, 165n51; agency-industry relations and, 56–63; costs of, 129; curriculum based on, 17; ecological forestry vs., 121–22, 131, 141; factors in, 39–41, 148n21; industrial forestry and, 63–66; market for, 66–67; and pricing, 41; science of, 15–17, 19; sustained-yield, 8, 10, 38; timber mining vs., 38; wildlife and, 94
World War II, 9, 56, 61, 77, 88, 97

Yale School of Forestry, 17
Yellowstone National Park, 4, 34, 159n72, 174n20
Yellowstone National Park Act (1894), 49
Yellowstone National Timberland Reserve, 34
Yosemite National Park, 43
Yosemite Valley, California, 4

zoning, 113, 170n114